# Switchgear Design, Operation, and Maintenance Using Industry Standards

# Switchgear Design, Operation, and Maintenance Using Industry Standards

## Protective Mechanisms, Sensing Technology, and Communication Standards


*Edited by*

**Gyan Ranjan Biswal**
Department of Electrical and Electronics Engineering,
Veer Surendra Sai University of Technology, Burla, Odisha, India

**Bhaveshkumar R. Bhalja**
Department of Electrical Engineering, Indian Institute of Technology
Roorkee, Roorkee, Uttrakhand, India



ELSEVIER

Elsevier
Radarweg 29, PO Box 211, 1000 AE Amsterdam, Netherlands
125 London Wall, London EC2Y 5AS, United Kingdom
50 Hampshire Street, 5th Floor, Cambridge, MA 02139, United States



ISBN: 978-0-443-24718-7

For information on all Elsevier publications visit our website at https://www.elsevier.com/books-and-journals

*Publisher:* Megan Ball
*Acquisitions Editor:* Rachel Pomery
*Editorial Project Manager:* Ellie Barnett
*Production Project Manager:* Anitha Sivaraj
*Cover Designer:* Christian Bilbow

Typeset by TNQ Tech

# Contents

# Part II
# Control and instrumentation and self-sustaining ecosystems

# Contributors

**Bhaveshkumar R. Bhalja**, Indian Institute of Technology Roorkee, Department of Electrical Engineering, Roorkee, Uttarakhand, India

**Gyan Ranjan Biswal**, Veer Surendra Sai University of Technology, Department of Electrical & Electronics Engineering, Burla, Odisha, India

**Sandhya Chandran**, Department of Electrical and Computer Engineering, University of Saskatchewan, Saskatoon, SK, Canada

**Sovan Das**, Kyocera SLD Laser Inc, Goleta, CA, United States

**Soumya Debashis Das**, Veer Surendra Sai University of Technology, Department of Electrical & Electronics Engineering, Burla, Odisha, India

**Kalyan Dash**, Continuum Associates LLC, Millburn, NJ, United States

**Vishal Kumar Gaur**, Motilal Nehru National Institute of Technology Allahabad, Electrical Engineering Department, Allahabad, Uttar Pradesh, India

**Ramakrishna Gokaraju**, Department of Electrical and Computer Engineering, University of Saskatchewan, Saskatoon, SK, Canada

**Tarikul Islam**, Jamia Millia Islamia (A Central University), Department of Electrical Engineering, Delhi, New Delhi, India

**Akhtar Kalam**, Victoria University, College of Engineering and Science, Melbourne, VIC, Australia

**Shailesh Kumar**, Arka Jain University, Department of Electrical and Electronics Engineering, Mohanpur, Jharkhand, India

**David Macdonald**, GE Vernova, Cambridge, MA, United States

**Rudra Prakash Maheshwari**, Indian Institute of Technology Roorkee, Roorkee, Uttarakhand, India

**Krishna Murari**, The University of Toledo, Department of Electrical Engineering & Computer Science, Toledo, OH, United States

**Vassilis C. Nikolaidis**, School of Electrical and Computer Engineering, National Technical University of Athens, Athens, Greece

**Balakrishna Pamulaparthy**, GE India, Hyderabad, Telangana, India

**Sweta Panda**, Veer Surendra Sai University of Technology, Department of Electrical & Electronics Engineering, Burla, Odisha, India

**Tarlochan Sidhu**, Ontario Tech University, Department of Electrical, Computer & Software Engineering, Oshawa, ON, Canada

**Manohar Singh**, Punjab Engineering College (Deemed to be University), Electrical Engineering Departmment, Chandigarh, India

**Biswajit Tewari**, Electrical & Instrumentation, Hindalco Industries Ltd., Hirakud, Odisha, India

# Foreword

Socioeconomic growth, rapid automation and the need to reduce greenhouse gases have given rise to an increase in the per-capita consumption of electricity around the globe. Modern electric power systems catering to huge energy demands are spread over wide areas and comprise numerous key components, such as switchgear and protective devices, designed to offer uninterrupted electrical power supply. The increase in demand has necessitated the use of large-capacity power equipment and complex interconnections among them. This has resulted in a multifold increase in pressure on the switchgear and protective devices.

During the evolution of the power system, a remarkable transition has been observed from small, isolated systems to wide-scale interconnected networks spread over multiple regions and countries, even across continents. The evolution also includes merging renewable energy sources with conventional sources. Although the evaluation has improved the observability and reliability of the system, it has also imposed significant challenges in protection, control and optimal power management. To address the said issues, engineers have integrated communication networks with the existing power infrastructure. Although the amalgamation of communication networks with power infrastructure resolves many issues, it also creates new challenges in the form of cyberattacks that hinder the safe and reliable operation of the power system.

The primary intention of this book edited by Gyan Ranjan Biswal and Bhaveshkumar R. Bhalja is to serve as a handbook for undergraduate and postgraduate students, as well as academics and researchers working at universities and research and development centers. Furthermore, this book can also be used as a reference and working book by engineers working in the industry.

This book starts with an introduction to switchgear in which the type and insulating medium used in the switchgear are discussed. Treatment of switchgear covers the concept of switchgear operation, industry-standard protective mechanisms, state-of-the-art sensing technology and relevant communication standards data links. This book is divided into three parts, namely, (1) Protection and Safety, (2) Control and Instrumentation and Self-Sustaining Ecosystem and (3) Operation and Maintenance.

In the first part, four chapters, protective mechanisms used in practice, switchgear faults and their remedies are described. Discussion related to DC-type circuit breakers is also given. No treatment of protection and safety can be complete without the introduction of design considerations of relay and circuit breakers.

Elaboration about control and instrumentation aspects of sub-station along with the self-sustaining ecosystem is given in the second part comprised of four chapters. Here, the control and instrumentation aspects of switchgear operation are given. Due importance is given to testing and validation of smart and intelligent sensors. Besides the control and instrumentation aspects, this book covers detailed theories of communication protocols used for substation automation. Cybersecurity aspects of the power grid and measures against cyber threats are also covered. Finally, the operation and maintenance of circuit breakers and high-voltage DC applications of switchgear are discussed in the third part comprised of two chapters.

In general, a thorough coverage of protection and safety, control and instrumentation, self-sustaining ecosystem and operation and maintenance of circuit breakers along with DC applications of circuit breakers is provided.

To summarize, this book will be an imperative source for beginners and practitioners in the field of protective mechanisms, sensing technology and communication standards.

**Prof. O. P. Malik**
*Professor Emeritus, University of Calgary, Calgary, AB, Canada*

# Preface

*Switchgear Design, Operation, and Maintenance using Industry Standards: Protective Mechanisms, Sensing Technology, and Communication Standards* is a practical handbook, equally useful for both industry experts and academicians covering the latest developments in switchgear.

This book breaks down cutting-edge practical techniques according to the hierarchy of switchgear operations, with an emphasis on critical technologies for automation in the energy transition. Following a helpful refresher on switchgear fundamentals, Part I examines essential safety considerations from fault identification and resolution to AC/DC-type circuit breakers and other protective mechanisms. Part II sets out operating principles and testing procedures for reliable smart sub-stations comprising smart/intelligent sensors and sensor networks, including communication protocols, validation and cybersecurity. Finally, Part III considers essential operation and maintenance practices being adapted as per the industry standards, such as circuit-breaker maintenance and the critical function of high-voltage DC switchgear for the energy transition.

An up-to-date helping hand for the transfer from university programs to industry, *Switchgear Design, Operation, and Maintenance using Industry Standards: Protective Mechanisms, Sensing Technology, and Communication Standards* will allow professionals to design, operate and maintain the smart, automated substations the energy transition needs.

Most of the available books in the market put their efforts into discussing the role of protection and the safety requirement of the switchgear which is one of the critical aspects of the entire operation of substations. However, in general, there are three hierarchies defined as per the industry standards referred to ensure the healthy functioning of substations. These are (1) Protection and Safety, (2) Control and Instrumentation and (3) Operation and Maintenance of the entire substation/switchgear.

In the first version of this book, the editors put their efforts into covering the existing and new challenges of switchgear safety and operation. It primarily covers the criticalities involved in the state-of-the-art protective mechanisms, sensing technology to deal with the sensory parameters and the modern communication standards to exchange the data between the circuit breaker and the human–machine interface (HMI)/control center of the switchgear. It also includes the required testing and calibration technology involved in both the AC and DC types' circuit breakers.

## Key features

- Tailors itself to industry standards and the practical hierarchy of switchgear operations for the maximum applications.
- Includes clear chapter objectives and case studies to support learning.
- Covers the latest switchgear developments for automated substations to support the energy transition.

# Acknowledgements

First and foremost, *Editors* would like to express their regards to the publisher for publishing state-of-the-art content on the switchgear operation and its applications which comprise recent developments. This book is compiled by contributions truly across the world from five continents right from academia to industry experts. We also thank the entire editorial team and publishing team for their suggestions, constructive criticism and encouragement throughout various phases of planning, initiation and, ultimately, completion of this study. Their advice helped us in a broad way for the completion of this unique book.

We express our sincere thanks to all the authors of this book; without them, it would not be imaginable to give such a unique shape to the handbook that will be equally effective to researchers, academicians and domain industry engineers for the decades to come. In this regard, we extend our thankfulness to the entire authorship, namely, Prof. Akhtar Kalam, *Victoria University, Sydney, Australia*; Prof. R. P. Maheshwari, *Retired, Indian Institute of Technology Roorkee, Uttarakhand, India*; Dr. Manohar Singh, *PEC University, Chandigarh, India*; Dr. Nikolaidis Vassilis, *Democritus University of Thrace, Greece*; Dr. Krishna Murari, *University of Toledo, Ohio, USA*; Dr. Vishal Kumar Gaur, *MN National Institute of Technology, Allahabad, India*; Prof. Ramakrishna Gokaraju and Sandhya Chandran, *University of Saskatchewan, Canada*; Dr. Sovan Das, *MTS Systems Engineer, Goleta, CA, USA*; Er. Soumya Debashis Das, *Veer Surendra Sai University of Technology (VSSUT), Burla, India*; Prof. Tarikul Islam, *Jamia Millia Islamia, New Delhi, India*; Dr. Shailesh Kumar, *Arka Jain University, Jharkhand, India*; Dr. David Macdonald, *GE Vernova, Cambridge, MA, USA*; Dr. Balakrishna Pamulaparthy, *GE India, Hyderabad, India*; Prof. T. S. Sidhu, *Ontario Tech University, Ontario, Canada*; Er. Sweta Panda, *VSSUT, Burla*; Er. Kalyan Dash, *Power System Engineer, Continuum Associates LLC, USA*; and Er. *Biswajit Tewari, Head E&I, HINDALCO, Aditya-Birla Group, India*, for their selfless and invaluable contributions. We also extend our thanks to Ms. Ellie Barnett, *Editorial Project Manager, and her entire team* for background support.

Finally, we dedicate this work to *mankind*.

**Gyan Ranjan Biswal and Bhaveshkumar R. Bhalja**

Chapter 1

# Introduction to switchgear: Types and insulation mediums

Akhtar Kalam[1], Rudra Prakash Maheshwari[2] and Gyan Ranjan Biswal[3]
[1]*Victoria University, College of Engineering and Science, Melbourne, VIC, Australia;* [2]*Indian Institute of Technology Roorkee, Roorkee, Uttarakhand, India;* [3]*Veer Surendra Sai University of Technology, Department of Electrical & Electronics Engineering, Burla, Odisha, India*

## 1. State of the art

The substations are one of the most crucial links in the electric power grid, and it has to be alive 24 cross-seven to keep the grid available to the local utility company that ultimately serves to power consumers. These substations are of two types that can be identified based on rated voltage level, and they are (1) transmission substations, typically vary between 220 kV and 765 kV and above, and (2) distribution substations that vary between 415 V and 33 kV as per the Indian Electricity Grid Code (IEGC), regulated by the Central Electricity Regulatory Commission (CERC) [1]. Therefore, types of insulation mediums are different, depending upon the capacity/rating of the substations. It can be air OR oil OR gas type even for equipment the insulation can be solid as well. Out of these, the gas-type insulation medium is the most popular one because of the distinct benefits offered by gas-insulated switchgear (GIS) as compared with its predecessors, and it is more popularly known as *transmission sub-stations*. The handbook has clearly emphasized the design and operational activities of these modern GISs. GIS and applications GIS is a vital component in the grid infrastructure that provides all the necessary interfaces to both the generating systems with the distribution systems. Therefore, the editors of the book have decided to cover the crucial technicalities involved in GIS operation as per the industry standards. It provides a unique blend of inputs to cover all three hierarchies of the functioning of substations, that is, (1) Protection and Safety; (2) Control and Instrumentation, and (3) Operation and Maintenance, as depicted in Fig. 1.1.

A special emphasis is placed on providing deep insights into protective mechanisms, sensing technology and current and upcoming communication standards/protocols to ensure reliable design, and safe operation and

**Switchgear Design, Operation, and Maintenance Using Industry Standards**
**https://doi.org/10.1016/B978-0-443-24718-7.00001-9**

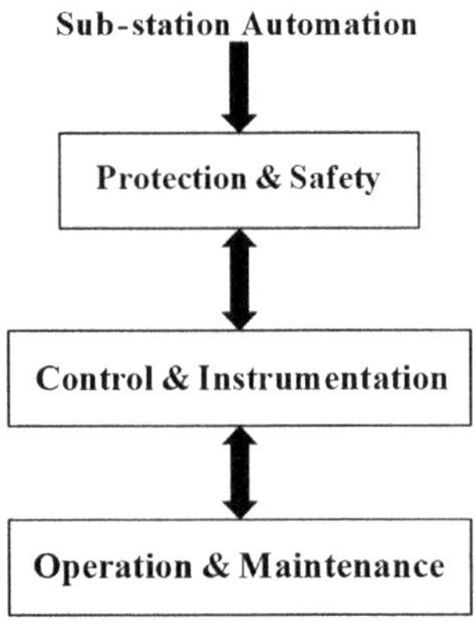


**FIGURE 1.1** Demonstrating operational hierarchy of substations.

maintenance (O&M) [2–8]. All the modern communication links are used for seamless data exchange between different sensors mounted on GIS and its human–machine interface (HMI)/computer centre (CC). It is to be noted that two terms are frequently used as synonyms of each other, that is, switchgear and substation throughout the handbook. The book has also thrown light on testing and calibration techniques that are suitable for both conventional high-voltage alternate current (HVAC) and new-age high-voltage direct current (HVDC) systems. That is why this handbook considers protecting interests between types of readership involved in academia (*academic institutions/universities/R&D laboratories*) and *industries* dealing with the operation and maintenance of GIS.

In this chapter, a major effort is given to primarily describe the type of insulation gases used in the present age switchgear ($SF_6$ as insulation medium) and futuristic mixed gases (N2:SF6::80:20) OR C3H7N (acetone amine) as an insulation medium in upcoming circuit breakers [7,11–13]. Smart grid technology necessitates the use of smarter components in the form of smart sensors and actuators to activate the protective mechanisms with the minimum response time. As per the industry practices and operational requirements, all the components of a GIS are installed inside the vessels of switchgear except the power transformers. Typical switchgear of 132-kV/33-kV size is demonstrated as shown in Fig. 1.2. Normally, pressures inside the vessels are kept above the atmospheric pressures and are varied from 1.9 Bar to 9–10 Bar, depending on types of design configurations introduced by companies such as ABB, GE and others [3,9]. However, as shown in Fig. 1.3, every one out of four failures is reported due to late detection of gas leakages from the GIS [3,8–12].

$SF_6$ gas is the most widely used in the present GIS because of its high dielectric constant, higher arc quenching capability and better heat dissipation as compared with air [5–7]. $SF_6$ as an insulation medium over the air is

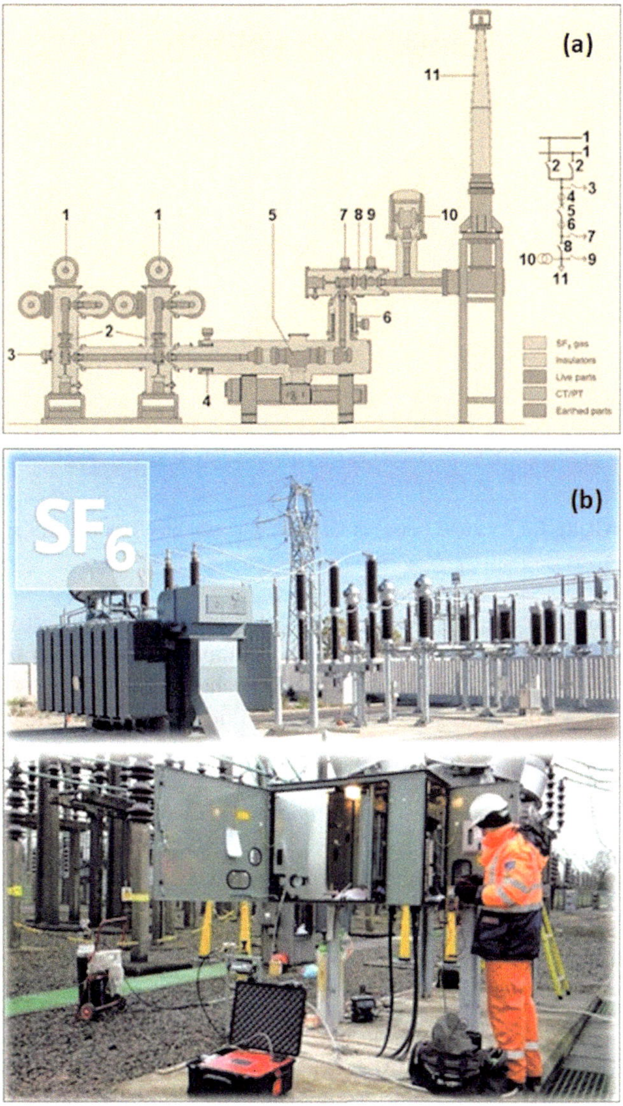


**FIGURE 1.2** (A) Piping and instrumentation diagram of a 132-kV size gas-insulated switchgear/transmission substation. (B) A filed demonstration of air-insulated switchgear.

preferred in GIS operation because of its enhanced specific heat and arc-quenching capabilities. It is an inert gas. However, a one-ton leakage of $SF_6$ causes 22,800 times more impact on the environment as compared with the same quantity of greenhouse gases (primarily $CO_2$) emissions. In particular, the presence of live conductors experiences sudden arcing at mechanically/

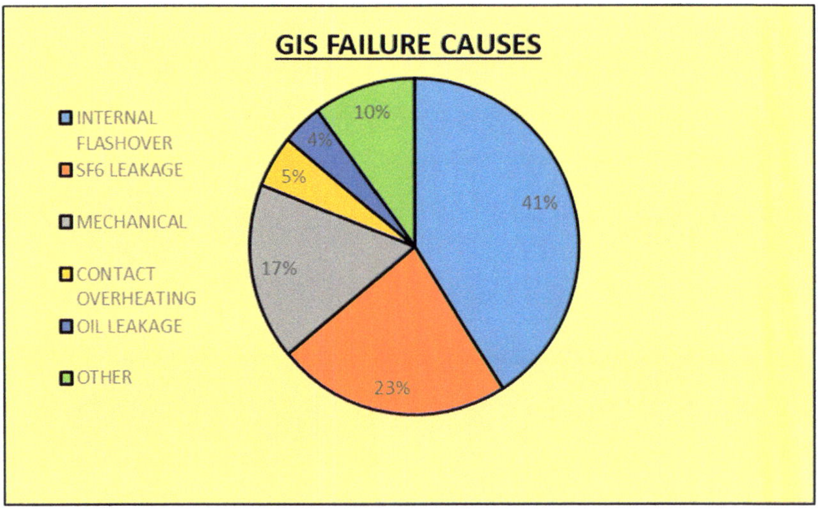


**FIGURE 1.3** Various causes of GIS failure in real-time conditions [3,8].

digitally coupled contact points inside GIS [7–9]. Now protective mechanisms/relays activate themselves to handle the anomalies whenever operational requirements deviate from the prescribed standards of O&M. A massive amount of energy is released during the switching operation. Arcing causes the sudden shoot-up or change in temperature at contact points in a highly pressurized environment. It is a reason why humidity appears for a transient period inside vessels in a *constrained environment (fixed volume)* due to thermodynamics applicable (*high pressure-low temperature to high pressure-high temperature, and back to high pressure-low temperature*). So, there are two major reasons that are as follows:

1. leakage quantity of insulating gas and
2. degradation in quality of insulating gas.

At present, the book is considering $SF_6$ as an insulation medium, which is the most popular one in the market. Further, the role of sensing technologies involved in the detection of gas leakage and humidity detection during arcing is elaborated in Section 1.3 (Fig. 1.2).

## 2. Types of switchgear and its insulation medium

As per the instrumentation standards defined for rated voltage, current, power and power factor by the *International Electro-technical Commission* (IEC) and the *Institute of Electrical and Electronics Engineers* (IEEE), switchgears and its insulation mediums can be categorized into three classes:

1. **Air**: Air is used as the primary insulation medium in many transmission substations due to its high dielectric strength and low cost. In most of the switchgear where the air is used as an insulating material, to decrease the impact of the arc, its resistance is increased by the blast of dry air. This blast elongates the arc length and thus arc resistance and quenching the arc. Another method that was initially very popular was creating a vacuum in the contact container, thus eliminating the medium to ionize. But with increasing voltage level, it became difficult to get the vacuum up to that required level. Moreover, the medium for heat dissipation was not available. But now with the development of technology, it is possible to get higher order vacuum, and vacuum circuit breakers are again being considered. Additional advantages of using air as a medium in switchgears include neatness, low maintenance demand, and are cleanliness.
2. **Oil**: Oil is used as an insulation medium in some transmission substations to provide higher dielectric strength and better heat dissipation than air. For this purpose, some organic oils were used initially, but now a mixture of organic and mineral oils is used. One of the challenges with oil is that due to high temperatures, oil gets decomposed and forms gas. This gas gets dissolved in oil adversely affecting its heat-carrying capacity. Yet another issue is that this gas causes an increase in volume so tank pressure may get elevated, or some space is required to be provided to accommodate this volumetric expansion. These types of switchgears require more mainte nance and are not very clean.
3. **Gas**: Certain gases, such as sulphur hexafluoride ($SF_6$) and pentane, are used as insulation mediums in transmission substations due to their low viscosity and good electrical properties. Switchgears using SF6 require more maintenance as this gas is required to be at positive pressure, thus eliminating any chance of oxygen entering the chamber. Yet another issue is that after every operation, some white powder is formed, which is required to be removed. Thus, their operational cost is also high. Yet another challenge is that these causes are due to the negative environmental impact of $SF_6$. Although $SF_6$ is very popular for medium- and high-voltage switchgears, its alternative is being searched, which is more environmentally friendly.

Besides these gas or liquid insulators for many applications, solid insulators are also required. These solid insulators are mostly used for outdoor applications. Keeping all parameters under consideration, ceramic is used mostly for fabrication of these insulators. To provide more dielectric strength and less weight, glass is also used for these applications. But due to the high cost of glass insulators, ceramic insulators are more popular until and unless the situation demands them. Based on applications, these insulators can also be classified. The following is their classification:

- **Pin-type insulators**: In pin insulators, a cavity is formed in the centre of the ceramic in which a steel pin is threaded. These types of insulators are used for distribution systems generally up to 33 kV. Due to the different coefficients of expansion of ceramic and steel, these develop fractures during extreme weather conditions. In these cracks, dust gets settled and during rains, which leads to flashing, which may precipitate solid faults. Based on operating voltages, the shape and size differ. In pin-type insulators, conductor is placed on the top and tied to it with a metallic wire.
- **Line post insulators**: As already described, these are also made using ceramic. The name comes from the application area. They are designed in a shape that they can be stacked to form a column. These are used to protect the charged element from getting grounded. The number of insulators is decided by the voltage rating of the equipment.
- **Strain insulators**: A strain insulator is an electrical insulator that is designed to work in mechanical tension (strain), to withstand the pull of a suspended electrical wire or cable. They are used in overhead electrical wiring, to support radio antennas and overhead power lines. Generally, they are made of ceramic, but in some cases where the force is much, glass is also used for it.
- **Shackle insulators**: Shackle insulators are one type of overhead insulator used to provide insulation to conductors on low-voltage power lines. It is also known as a spool insulator or a butterfly insulator. They are typically installed horizontally or vertically. In this type of insulator conductor is wrapped around it. A hole is there in the centre trough which a pin passes to mount it on a pole through a clamp.
- **Post insulators**: Postinsulators are used as supports for bus-bars in transformer substation yards, in switchgear as support for breakers, and for capacitor banks. These insulators are generally subjected to bending, torsion and compression forces in service. As compared with pin insulators, they have a bigger profile.
- **Cap and pin insulators**: Cap and pin insulators of porcelain or glass predominate as overhead line suspension or tension sets above 33 kV, and they are also used for substation busbar high-level strained connections. Almost any creepage distance may be achieved by arranging the required number of individual units in a string. They are designed such that a single disk can be replaced and several disks can be varied based on the voltage level.
- **Stay insulators**: Stay insulator is a type of low-voltage insulator designed to fasten and counterweight the dead-end pole by connecting with a stay wire or guy grip. Owing to the reason they are used to fasten the counterweights, they have very good mechanical strength. These are used in distribution systems.
- **Hollow insulators**: Hollow insulators are components of the bushings that ensure the safe handling of electricity by transformers and circuit breakers.

They are also used in instrument transformers, which measure voltage and current. Being their centre hollow, they are called hollow insulators. This hole provides an insulated path for a live conductor connected to the device.

## 3. Role of smart sensors to make GIS smarter

Before understanding the role of a smart sensor in pursuing the measurement of parameters/variables, we need to brainstorm the following questionnaires:

- (Q1) What is smart sensor?
- (Q2) Is smart sensor different than an intelligent sensor?
- (Q3) Is smart sensor different than a conventional physical/chemical sensor?
- (Q4) Are terms such as transducer, sensor and actuator different or indifferent?
- (Q5) Are sensor(s)/actuator(s) integral part of any automation system?

Let us find the answers to the aforementioned points, and develop our basics clearly before moving ahead to understand the *role of smart sensors to make GIS smarter.* Thinking about it, Fig. 1.4.

Substation automation can be defined as *the act of controlling the substation(s)/load dispatch centre (LDC)/switchgear via automated processes within computers and intelligent C&I devices* using an integrated power system [14]. The engineers and designers of industries have always tried to automate the processes/plants as much as possible, based on available instruments. There has been a continuous development in industrial processes, which needed the development of better instrumentation and automatic control.

**FIGURE 1.4** Think over it.

## 3.1 Smart sensors

A module containing sensor element(s) suitability is integrated with necessary electronics such that the output is truly or easily compatible with the intended end-device, and the module usually takes the form of a single IC (integrated circuit) chip, which is known as 'smart sensor'. Here, a digital processing unit (DPU) may or may not be part of signal-conditioning (SC) circuits or it can be added to the SC circuit. * *Here, we find an answer to our first* **Q1**. Experts may still prefer to debate this topic and try to come up with their definition. However, all the terminologies are defined here, adapted from the IEEE Society of Instrumentation & Measurement, and library of recorded research findings articulated on time to time basis across top research lead universities in the world. Moving forward, the terms 'smart' and 'intelligent' are again relative and debatable. Thus, let us try to define both terms to satisfy the minimum requirements, while we are expecting an optimal outcome from the automation system, which is in practice to control and monitor the entire station/plant/process. So, now let us fix the definition of 'intelligent sensor' first, and then we shall be in a position to conduct a qualitative comparison between the two terms.

A module containing sensor element(s) suitability is integrated with necessary electronics that essentially includes DPU and some features of human–machine interactions such that the output is truly or easily compatible with the intended end-device, and the module usually takes the form of a single IC (integrated circuit) chip, which is known as 'intelligent sensor'. A DPU can be a type of (1) microprocessor, (2) microcontroller or (3) digital signal processor including required memory and I/O devices. The human–machine interactions and embedding of the decision-making algorithms are the relative level of intelligence that may or may not be a part of solutions being offered for the accomplishment of a particular task and again is *debatable*. * *Here, we find an answer to our first* **Q2**. Now you can consider adding DPU as the last block in Fig. 1.5, and fetching decision algorithms to bring the human–machine interaction into reality by considering it as the last block of Fig. 1.6 to satisfy the current consciousness among academic and research fraternities.

Yes, a smart sensor is the next-generation sensor when we compare it with a conventional form of sensor called physical sensor/chemical sensor.

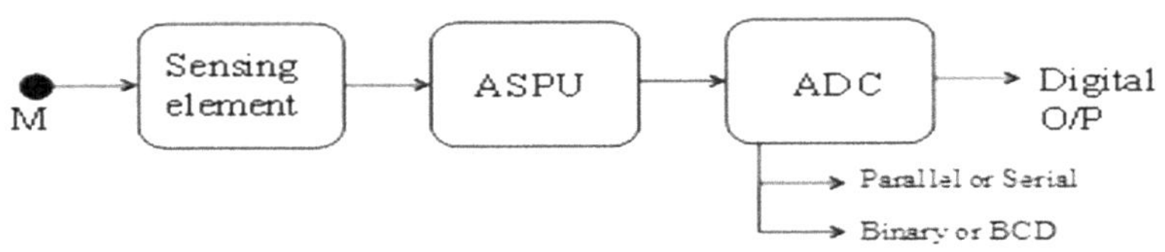


**FIGURE 1.5** Schematic diagram representing level II type of smart sensor.

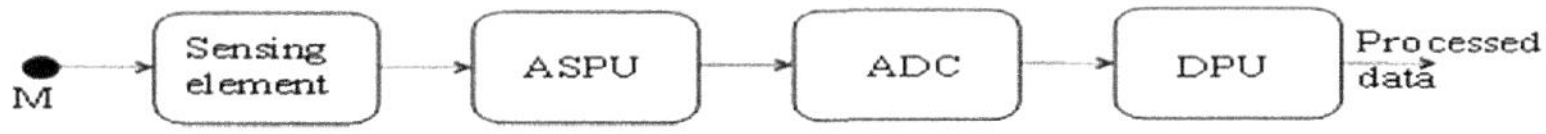


**FIGURE 1.6** Schematic showing level III type of intelligent sensor.

Primarily, one of the ways to classify a conventional sensor is that it can be either recognized as a physical/mechanical sensor or it can be type of chemical sensor. When we recognize a device as a sensor, then it is expected that a sensor follows some principle of transduction (which is another way to classify sensors). Typically, the final output of the sensor shall be available in electrical form, that is, relative voltage or relative current, and therefore, the final electrical form of data can be easily memorized/stored/displayed/processed using an in-built processor. It is not at all possible to memorize/store/display/process data in any other form such as physical/chemical other than the electrical form of data as most of the signal conditioning devices are electronic circuits. **I am sure after referring to the statements*; *readers will be able to appreciate the possible answer for* **Q3**. In this regard, a basic schematic diagram of the host−computer instrumentation system is shown in Fig. 1.7. Primarily, a smart sensor is fabricated by merging two critical technologies, that is, micromachining/nano-machining and microelectronics/nano-electronics, depending on design, fabrication, cost and applications to be fabricated a sensor at the micron/the nanoscale.

Now, that we have developed our understanding of a smart/intelligent sensor that meets the current IEC/IEEE standards, let us explore *how and why* the domain experts and industries use terms such as transducer, sensor and actuator in synonyms. Are these three terms interrelated or isolated or share some bonding? This exercise will help us to find the solutions for **Q4**. This symbolic key as shown in Fig. 1.8 reflects how and why?

As per the Instrumentation Society of America, a sensor/transducer is a device that provides a useable output in response to a specified 'measurand'. Here, measurand is the physical quantity/property/condition which is to be measured, and output is defined as an electrical quantity [17]. While; a

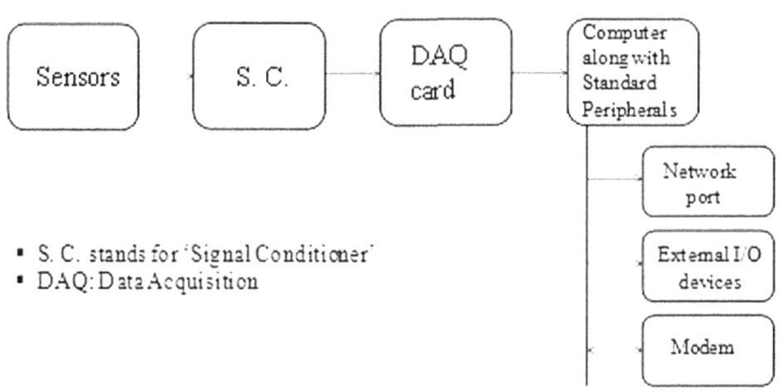


**FIGURE 1.7** Schematic diagram showing the host−computer instrumentation system.

**FIGURE 1.8** How and why?

physical quantity can be pressure, heat, flow rate, liquid level, flow, position, humidity, etc., and the response/output is measured in terms of parametric change or electrical signal. However, this does not bring clarity to **Q4**.

***Let us revisit to the basics↓**

A *transducer* is a device that works on the principle of transduction, which converts one form of energy into another form for measurement/control, whereas a *sensor* comes as a transducer along with necessary electrodes and input/output (I/O) mechanisms so that the final response of a sensor will be truly or nearly compatible with the intended end device of the application. On the other hand, an *actuator* is used to convert the electrical signal to a physical signal/variable, which is finally attached to some applications. In simplest form, one can say that both sensor and actuator come under the umbrella of transducer, where a *sensor* can be treated as an *input transducer* while an *actuator* can be treated as an *inverse or output transducer.* The salient points depicted in the following establish the relationship among the transducer, sensor and actuator [17]. * *Here, we find an answer to our* first **Q4**.

- Transducer: device that converts one form of energy into another form for the purpose of measurement/control.
- Sensor: Transducer (physical into electrical) + electrodes and I/O mechanisms.
- A sensor is an I/P transducer.
- An actuator is an O/P transducer.

Fig. 1.9 shows a closed-loop block diagram of an industrial measurement and control model where, designated components are participating to achieve a desired output. The model contains types of machinery (power station/plant/factory/electrical grid), controller(s) (primarily digital controllers for example, programmable automation controller/programmable logic controller), a human−machine interface (HMI), external or natural disturbances (noise), analog-to-digital converter (ADC) or digital-to-analog converter (DAC), and measurement systems (sensors, actuators). The figure shows interrelations among all the components that are participating in the feed-forward or feedback line(s) of an automatic control and monitoring system.

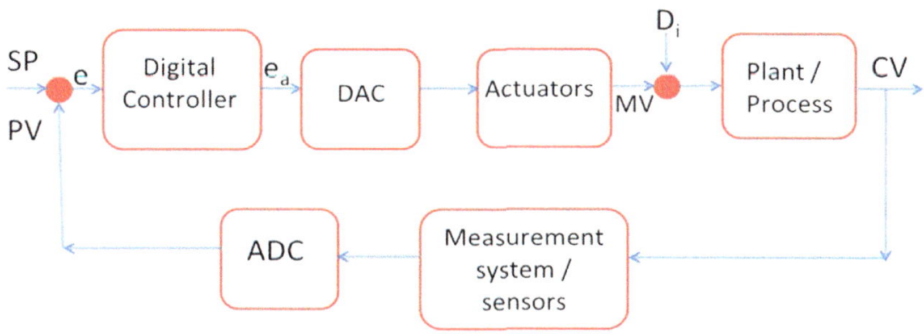


**FIGURE 1.9** An equivalent schematic diagram of measurement and control system for industry applications [17].

Now, we shall try to find a possible answer to the final point, which is **Q5** [Are sensor(s)/actuator(s) an integral part of any automation system?]. As per the standard model of automation defined at the beginning of Section 1.3, the terms control and instrumentation (C&I)/instrumentation and control (I&C) are used as synonyms of *automation*. It is to be noted here that broadly C&I or I&C have the same meaning and are used by industry experts and researchers frequently the way they are comfortable with. Out of the C&I, *control* is the *front-end* system, while *instrumentation* is the *back-end* system for any commercial/defence/scientific applications. The instrumentation system architecture acts in the *front end*, whereas the dedicated control system acts in the *back end* as far as industry applications are concerned. Typically, specialized areas such as control systems and instrumentation engineering are application-dependent, and they cannot be fixed as a model for every possible application. Rather it changes depending on applications. As shown in Fig. 1.9, controller(s), sensor(s) and actuator(s), all participate as a component of the presented closed-loop model. On that point, it is concluded that *sensor(s)/actuator(s) are an integral part of any automation system. * It is expected that readers find their answer of* **Q5**. Having a clear understanding of sensing technology, Q1−Q5, we are digging deep into the role of smart sensors in GIS applications. For that, we need to understand about reason for failures as depicted in Fig. 1.3.

### 3.2 Sensing technology in gas-insulated switchgear

Here, we have considered $SF_6$-GIS as a part of a case study, which is the most reliable type of transmission substations in the 21st century. The *degradation in $SF_6$ density* (because of leakage) and in *$SF_6$ quality* (due to moisture) directly affects the insulation quality and reliability of gas-insulated switchgear [7,13−16]. Two possible types of measurement systems can be offered by implementing a real-time automation system as per the IEC/IEEE Standards such as IEC 61850/61851/IEEE Std. C37.122.1. It is either an indirect measurement technique [8,11,12] or a new age state-of-the-art direct measurement technique [7,14–16].

Two types of SF6 density transducers are currently commercially available, and they can be retrofitted to a circuit breaker for $SF_6$ density monitoring. The first type of density transducer measures the pressure and internal SF6 gas temperature to derive density [18]. The gas temperature is typically recorded where the sensor is installed, which can affect the accuracy of the measurement due to heat flux. The second type of transducer available uses the quartz oscillating principle to directly measure density [19]. The quartz oscillating principle compares the relationship between dynamic viscosity and density of a specific gas at a specific temperature [20]. Individual pressure and temperature sensors can also be purchased to measure for SF6 leaks, but the measurements must be entered into an equation to be expressed as a density value. For this system to work correctly, the temperature sensor needs to provide an accurate representation of the ambient conditions. Therefore, mounting the temperature sensor in the right position is crucial if this setup is used. Here, we are considering humidity measurement as a by-product of the recent findings reported in Ref. [21].

To conduct the indirect density measurement in $SF_6$-GIS, the sensors commercially available are based on resonant frequency shift of quartz tuning fork placed in $SF_6$ medium. The quartz tuning fork sensors operate on the principle that the added mass of the fluid surrounding the resonating device causes a shift in the resonating frequency, which can be conveniently measured [22]. Since the output of these sensors depends on the damping effect on the vibrating mass, small variations in mass due to manufacturing differences limit their accuracy. These limitations are currently overcome by calibration and curve-fitting using empirical methods [23]. It has also come with another challenge that these tuning fork methods are originally designed for measuring the relative change in density in a liquid environment. It succeeds because per unit volume of a liquid is much higher/denser as compared with the same unit of gas. However, the same technique has been proven to be inaccurate when it comes to the detection of minute amounts of gas leakage. *It can detect only if a substantial amount of gas leaks into the atmosphere, and that's the biggest challenge.* It is the reason why these indirect methods of gas leak detection failed numbers of times, and is a major cause of GIS failures [7,8,16,17].

Moisture in $SF_6$ gas is another important parameter, which deteriorates the insulation property of the gas, thus causing failure of the substation. Moisture measurement in $SF_6$ gas needs special attention of low concentration measurement, which is always challenging by the conventional sensor [18–21]. Metal oxide thin film capacitive sensor, which is thermally and chemically stable, may be suitable for the low ppm moisture measurement in $SF_6$ gas. It has been observed from the recent findings that a capacitive sensor is suitable for traces of moisture measurement in SF6 gas. The sensor has been fabricated using fluorine-doped tin oxide (FTO) glass as the base substrate and silver as the upper electrode. A thin film of porous alumina ($Al_2O_3$) was used as a sensing layer for detecting the presence of moisture at parts-per-million (ppm) scale. The performance of the presented sensor was experimentally validated for different moisture concentrations in SF6 gas. The sensor shows a good sensitivity (0.238 pF/ppm) with linear characteristics (±2.05%) over the moisture range of 246–725 ppm. The response of the fabricated sensor is very accurate (±0.84%), with low hysteresis (±0.32%) and highly repeatable (±0.013%) [15]. Simultaneously, an RH-scale (relative humidity) smart humidity sensor was also fabricated using poly-vinyl alcohol (PVA) as sensing film deposited on a polyimide substrate. This sensor detects changes in capacitance due to the presence of humidity in GIS applications. The initial tests are conducted using moist nitrogen gas with different percentages relative (% RH). The performance of the designed PVA sensor was experimentally validated, and it was noticed that the sensor possesses good sensitivity with linear characteristics over the range of 0%–20% suitable for industrial practices [16]. Both are highly reliable designs as both the sensing layer and substrate form a cohesive bond between each other as both materials are organic. One offers a highly linear response at ppm-scale, whereas the other one offers detection in % RH scale.

## 4. Operation and maintenance of the GIS: Automation for real-time operation, preventive maintenance and fault diagnosis of GIS

Operation and maintenance of a plant/power station/grid/factory is essential to attain its complete life cycle. Instead of that, the overall life cycle of a GIS (in this case) may substantially reduce; especially, the *useful life-period* of the GIS/substation [14]. In the present scenario, components of substations may be influenced by three factors that change the performance significantly: accuracy, efficiency and reliability. After the commissioning of highly accurate digital controllers, system reliability is the main impact index for its performance quality. The designer's prime objective is to carry out the process with utmost efficiency of the system having maximum availability at minimum cost that also ensures uninterrupted power supply. Here, three important aspects need to be touched on for the smooth operation and maintenance of a GIS.

1. Real-time data acquisition, fault diagnosis and monitoring of parameters/variables
2. Preventive maintenance of all components of the station

Complex systems require monitoring and diagnosis by operating personnel as well as excellent automation. Now, point (1) can be attained by implementing an automation platform that ensures that the time-stamping of data (sensed, collected, reported to CC) must be less than the response time of protective relays activated to avoid abnormal situations. It is to be noted that protective relays always operate based on feedback provided by sensors mounted in the field to detect possible types of faults and convey these data to CC. Such coordination between the P&S section and the C&I section is very important to ensure a stable operation of the dynamic electrical power grid. In place of that, a substation entire grid may collapse and black out the grid [1,3,14].

At this stage, it is assumed that the automation platform has been chosen by the planners to control and monitor the grid embedded in all the essential features as per the international standards such as IEEE C37.1.x/IEEE 1451.5 Std [24,25]. Phasor-measurement units (PMUs) and conventional supervisory control and data acquisition (SCADA) systems (those were introduced a decade back) are different as one uses phase-shift keying (PSK) to time-stamp its data, while the predecessor uses amplitude shift keying (ASK) to time-stamp data at the time of sensing and detecting faults. PMUs offer faster operational advantages by detecting faults early. So, it needs to be clear that the present-age SCADA has embedded all the benefits of PMUs, and both the units work in tandem, depending upon the design requirements of an integrated grid. For example, in India (Bharat), all the regional grids/load dispatch centres are fully integrated, and thus, decisions on localizations of PMUs and SCADA are taken accordingly, and it can be upgraded in the future, if necessary. These digital automation systems execute their tasks based on predefined algorithms up to 99.99% accuracy. So, the accuracy of operations has not been a major concern since the introduction of the automation system. Efficient operations of individual components are discussed later in Chapters 10 and 11 under the O&M section.

*It is important to share how you see a PMU/SCADA or any automation platform as a system.* Conceptually, an automation platform is the combination of two terms, namely, 'telemetry' and 'remote control'. When measurand (M) and user (in case of industrial automation, operator/system engineer) are separated in such a way that local measurement methods are unable to measure the measuring quantities, viz., current, voltage, power, power factor, flow, level, pressure, temperature, moisture concentration, etc. due to distance and/or inaccessibility, it needs some method of electrical communication scheme called '*telemetry*'. Temperatures inside the vessels of GIS and of the transformer core and pressure inside the closed chamber are some examples where it requires applying some method of telemetry. Nowadays, digital telemetry schemes have replaced D.C. telemetry systems and A.C. telemetry systems because of (1) error detection and correction capabilities, (2) measuring a large number of physical quantities and (3) improved reliability, except in some short-range applications where A.C. telemetry has proven cheaper, whereas pulse modulation schemes such as pulse code modulation (PCM), pulse width modulation (PWM) and pulse position modulation (PPM) are used to implement time-division multiplexing (TDM)/frequency-division multiplexing (FDM) systems, by which multiple RTUs can interact with CC (control centre). *Remote control means control of remote objects not control of variables* (voltage, current, pressure, etc.) as in the case of telemetry. Thus, reliability is the most important criterion for the success of a remote control system, and hence, it is used to deal with binary messages. If both telemetry and remote control merge, then the system is recognized as an automation system that is specifically designed and commissioned to meet predefined objectives. The major functions of automation platform are as follows:

- Data acquisition
- Monitoring
- Control
- Data collection and storage, calculations
- Report generation

Further, types of items in controlled plant and their functions are as follows in Table 1.1:

**TABLE 1.1 Types of functions of an automation system.**

| S. No. | Function type | | Data acquisition |
|---|---|---|---|
| 1. | Controlled variables | : | Acquire the analog values |
| 2. | Uncontrolled variables | : | Acquire the analog values |
| 3. | Remotely controlled objects | : | Acquire the status information/discrete values |
| 4. | Locally controlled objects | : | Acquire the status information/discrete values |

Preventive maintenance is the second factor to avoid any catastrophic failure of the GIS. Therefore, *point number (2)* can be attained by analyzing the system reliability of the entire GIS/substation, which is a married of numbers of components. At this stage, the single-line diagram (SLD) and the piping-and-instrumentation diagram (P&ID) are essential to conduct system reliability test and fault tree analyses for designing the GIS [26–28]. Typically, whenever term reliability is used, it refers to component reliability; not the system reliability. To provide a required level of system reliability, redundant elements are included. Usually, engineers try to achieve this level with minimal cost. The problem of total investment cost minimization, subject to reliability constraints, is well known as the redundancy optimization problem. The redundancy optimization problem for a multi-state system, which may consist of elements with different performance rates and reliability, is addressed as a problem of system reliability optimization (SRO). As depicted in Fig. 1.10, SRO is backed by bidirectional/two stage analyses

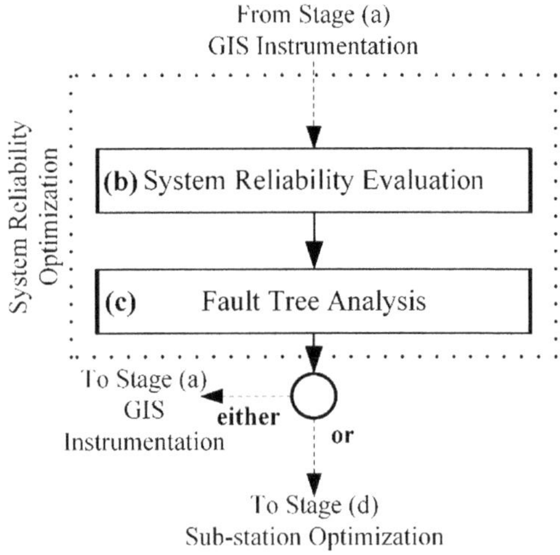


**FIGURE 1.10** The schematic diagram representing 'system reliability optimization' [14].

approach, namely, system reliability evaluation and fault tree analysis (FTA) in the scheme.

In general, fault avoidance methods prove to be less expensive than fault tolerance methods to evaluate system reliability. The fault avoidance methods and fault tolerance methods are used to evaluate the modal life of components and redundancy-based improvement in system design, respectively. In case of fault avoidance methods, the cost increases exponentially to a linear improvement in reliability performance of the system. Also, it is observed that the behaviour of components used as a part of the system can be different than the characteristics of components used as an individual. Thus, in industrial applications, one of the fault tolerance-based methods is preferred to evaluate the system reliability of the present and the already-existing systems [25,26]. Fault tree analysis (FTA) is a logical and structured process that can help to identify potential causes of system failure before the failures actually occur. Evaluation of the chances of system failure is one of the key approaches to test the system reliability/availability of the plant processes. FTA is a powerful design tool that can help to ensure that the objective performances of the model are met. It is a top-down, deductive analytical method for evaluating the chances of system failure. It is also a highly efficient method to study the behaviour of process models at supervisory/plant level.

Further, the book extensively covers the following features of switchgear:

- Protection and safety [6,10]
- Control and instrumentation standards and architectures [7,8]
- Operation and control practices [2,3]
- Testing and calibrations [2]

- DC-type circuit breakers: A new age technology in substation automation [4,9]
- Controlled switching applications of circuit breakers [5,9]

The first three bulleted points state about the three-level of hierarchies of a GIS as per the industry that detects the responsibilities of highly skilled engineers and personnel involved in day-to-day operations of the switchgear, while the rest three from bottom touch the critical aspects of GIS. The aforementioned six aspects are described under 11 different chapters in a systematic way by the academicians and industry experts whom are actively engaged in R&D and O&M of switchgear and its applications.

## 5. Organization of the book

The handbook has been organized into 11 chapters, and the essence of each chapter has been presented in the following sequence:

Chapter 1 sets forth the motive behind writing a book on switchgear operations as per the industry standards into three hierarchies, namely, (1) Protection and Safety, (2) Control and Instrumentation and (3) Operation and Maintenance. The book has given special attention to draughting content equally important for academic purposes and industry applications. It also covers recent advances in digital protection, safety standards, control algorithms, smart and intelligent sensors, wireless sensor networks, communication protocols and HVDC technology.

The description of the P&S (Protection and Safety) aspects of GIS, the first in the hierarchy, begins with Chapter 2. Primarily, the chapter comprises knowledge of protection practices in power utilities. It focuses on protection infrastructure which is essential to safeguard utilities against faults.

Chapter 3 brings out about DC-type circuit breakers in the book. It describes the role of a DC circuit breaker which is used to interrupt a DC network to tackle any possible chances of abnormalities under faulty or emergency conditions. In continuation to that Chapter 4 elaborates different possibilities of switchgear faults and remedies. This chapter provides in-depth exposure to switchgear faults that can lead to interruptions in power supply and related safety hazards. Finally, design considerations of relay and circuit breakers are covered in Chapter 5.

The C&I (Control and Instrumentation) comes into picture immediately next to the Protection & Safety hierarchy which is used to automate the control and monitoring activities involved in operating the GIS. The concept of C&I begins with Chapter 6. The chapter emphasizes on necessary technology and tools that regulate, monitor and opt for necessary decision-making algorithms to safeguard the electrical network.

The design, testing and validation of smart/intelligent sensors are elaborated in Chapter 7. In this chapter, all the sensors are primarily covered under two categories, that is, to monitor physical parameters and electrical parameters/variables. However, the authors have given special emphasis on physical parameters that are given limited exposures in available content in the same segment.

Chapter 8 describes the role of communication protocols in substation automation and its utility for enhancing the performance supervision of GIS in real time. Different WSN protocols are developed and available in the market to ensure the efficient operation, monitoring and safeguarding of the critical equipment of transmission substations.

Chapter 9 presents novel approaches to cybersecurity aspects of the book which is one of the most challenging and much-sought research domains. This chapter elaborates on the use of cybersecurity technology to keep safe protective equipment from possible chances of cyberthreats. It also gives exposure to the amalgamation of communication networks with power infrastructure challenges that hinder the safe and reliable operation of electrical circuits.

Finally, this handbook introduces the O&M (Operation and Maintenance) procedures of GIS through Chapter 10, which are the third and last in the hierarchy per industry standards. This chapter explores important topics such as the circuit breaker phenomenon and condition monitoring of different parameters of circuit breakers. Further, Chapter 11 examines the critical aspects of HVDC switchgear, focussing on its significance in modern power systems. Overall, this chapter provides valuable insights into the technological advancements, regulatory compliance and environmental considerations of HVDC switchgear. Both the chapters related to O&M aspects are written under the active observations of industry experts.

Necessary 'references' are cited by the authors at the end of the chapter as and when required to satisfy publication requirements. Different case studies are also presented to develop a better understanding of the domain knowledge mechanisms.

## References


[1] The Indian electricity grid code with vide. No. L-1/265/2022/CERC. June 07, 2022. https://cercind.gov.in/2022/draft_reg/Draft-IEGC-07062022.pdf.

[2] M. Kamei, O. Takai, Influence of sensor information accuracy on condition-based maintenance strategy for GIS/GCB maintenance, IEEE Trans. Power Deliv. 26 (2) (April 2011) 625–631.

[3] Interoperability for Advanced Protection and Control Applications, Standard IEC 61850 Protocol, 2019. https://www.se.com/ww/en/product-range-presentation/60793-iec-61850-standard/.

[4] Industries and applications of HVAC. Construction Materials and Artificial Monitoring, Helsinki, Finland. https://www.vaisala.com/en.

[5] J. Liu, G.M. Huang, Z. Ma, Y. Geng, A novel smart high-voltage circuit breaker for smart grid applications, IEEE Trans. Smart Grid 2 (2) (June 2011) 254–264.

[6] B. Bhalja, R.P. Maheshwari, N.G. Chothani, Types of circuit breaker and their testing, in: Protection and Switchgear, first ed., Oxford Univ. Press, New Delhi, India, 2013, pp. 344–387 (Chapter 14).

[7] G. R. Biswal, T. Islam, S. D. Das, and K. J. Akram, "Ultrasonic sensor based gas density monitoring of SF6 gas insulated switchgear," Indian Patent Application No.: 202111011524 A, Filing Date: Mar. 18, 2021; Publication Date: Apr. 16, 2021, Awarded Date: Jan. 29, 2024 (for 20 Years from 2021—2041).

[8] D. Zeisel, H. Menzi, L. Ullrich, A precise and robust quartz sensor based on tuning forl technology for (SF6)-gas desnity control, Sensor. Actuator. 80 (2000) 233–236.

[9] IEEE Standard for Intelligent Electronic Devices Cyber Security Capabilities, IEEE Standard 1686-2013, January 2014, pp. 1–29.

[10] M. Abdollah, A.A. Razi-Kazemi, Intelligent failure diagnosis for gas circuit breakers based on dynamic resistance measurements, IEEE Trans. Instrum. Meas. 68 (9) (Sep. 2019) 3066–3077.

[11] GE Sensing and Inspection Technologies, Moisture image series 1, datasheet, (2005) Available: https://www.gesensinginspection.com/.

[12] COSA Xentaur Corporation, XPDM, datasheet, (2020) Available: https://www.cosaxentaur.com/.

[13] IEEE guide for moisture measurement and control in $SF_6$ gas insulated equipment, in: Standard C37.122.5-2013, December 2013, pp. 1–42.

[14] G.R. Biswal, "PhD Thesis: Optimal HMI of Hydrogen Cooling System for Large Generators with Enhanced System Reliability", under the Supervision of Prof. R. P. Maheshwari and Prof. M. L. Dewal at Indian Institute of Technology Roorkee, Roorkee, November 2012.

[15] Z.H. Zargar, K.J. Akram, G.R. Biswal, T. Islam, A linear capacitive sensor for PPM moisture measurement in $SF_6$ gas insulated switchgear, IEEE Trans. Instrum. Meas. 70 (January 2021) 1–8.

[16] G.R. Biswal, P. Mohanty, K.J. Akram, N.P. Padhy, T. Islam, Design and fabrication of an inexpensive capacitive humidity sensor for smart sub-station automation, IEEE Sensor. J. 20 (12) (June 2020) 6215–6223.

[17] You-Tube Lecture Series on, Measurement and Instrumentation, Module 5, 2021, authored by G.R. Biswal. https://www.youtube.com/channel/UC0ISZ4dMZcIBeIzjZVRZhJw?view_as=subscriber.

[18] P. Cheetham, A. Hellany, S. Jones, Density monitoring of high-voltage $SF_6$ circuit breakers, IEEE Electr. Insul. Mag. 31 (2) (2015) 6–13.

[19] R. Rosario, R. Mutharasan, Piezoelectric excited millimeter sized cantilever sensors for measuring gas density changes, Sensor. Actuator. B 192 (2014) 99–104.

[20] T. Manzaneque, et al., Piezoelectric MEMS resonator-based oscillator for density and viscosity sensing, Sensor. Actuator. 220 (2014) 305–315.

[21] O. Cakmak, E. Ermek, N. Kilinc, G.G. Yaralioglu, H. Urey, Precision density measurement using two cantilevers with different widths, Sensor. Actuator. 232 (2015) 141–147.

[22] Rhodes, et al., Method and apparatus for monitoring $SF^6$ gas and electric utility apparatus, US Patent 2007/0027640 A1, 2007.

[23] A. Kramer, T.A. Paul, High-precision density sensor for concentration monitoring of binary gas mixtures, Sensor. Actuator. 202 (2013) 52–56.

[24] J.J. Arnold, J.R. Capener, Modern power station practice, in: Incorporating Modern Power System Practice, third ed.Turbine, Generators and Associated Plant, vol. 3, British Electricity International, London, U.K., 2011 ch. 1-12.

[25] G.R. Biswal, R.P. Maheshwari, M.L. Dewal, Modeling, control and monitoring of S3RS based hydrogen cooling system in thermal power plant, IEEE Trans. Ind. Electron. 59 (1) (Jan. 2012) 562–570.

[26] A. Birolini, Reliability engineering, Theor. Pract. (2007) 57–260.

[27] K.H. Wang, W.L. Dong, J.B. Ke, Comparison of reliability and the availability between four systems with warm standby components and standby switching failures, J. Appl. Math. Comput. 183 (2006) 1310–1322.

[28] G. Levitin, K. Hausken, Redundancy vs. protection in defending parallel systems against unintentional and intentional impacts, IEEE Trans. Reliab. 58 (4) (December 2009) 679–690.

# Part I

# Protection and safety

Chapter 2

# Protective mechanisms

Manohar Singh
*Punjab Engineering College (Deemed to be University), Electrical Engineering Departmment, Chandigarh, India*

## 1. Introduction

The operation of power components within their rated value tolerances is crucial for maintaining the stability and reliability of power systems. However, during electrical faults or abnormal operating conditions, these components can be driven outside their normal range, posing risks to the overall system health [1]. Effective power systems protection is essential to detect and swiftly clear faults, isolating faulty components and preventing widespread disruptions. Primary protection mechanisms are specifically designed to interrupt the flow of heavy short-circuit currents, thereby averting thermal damage to equipment caused by faults.

Different types of power utilities have distinct protection requirements. Power distribution utilities primarily deal with power distribution transformers, 66/33/11-kV feeders, and sometimes fixed capacitor banks [2]. Their primary concern is to protect distribution transformers during faults and restore power supply promptly [3]. Therefore, dedicated protection measures are implemented to safeguard transformers against short-circuit damage, supplemented by current backup protection to isolate transformers during through faults [4].

In contrast, transmission utilities manage interconnected power networks comprising power transformers, long transmission lines, bus bars and reactors. The high interconnectivity of transmission systems necessitates a focus on system stability to prevent thermal damage to interconnected power transformers, among other challenges.

Generating units serve as the primary source of short circuits [5], and their rotational inertia makes them particularly susceptible to mechanical, thermal and electrical damage. Hence, fast protection schemes are crucial to safeguard generating units from harm while maintaining synchronism.

**Switchgear Design, Operation, and Maintenance Using Industry Standards**
**https://doi.org/10.1016/B978-0-443-24718-7.00002-0**

The integration of solar and wind-based power generating resources introduces new challenges due to their rich power electronics interface, which limits their fault-feeding capabilities [6]. While these resources are less prone to thermal damage and synchronism issues, detecting faults remains a significant challenge [7]. Consequently, prioritizing power reliability over thermal damage and system stability is paramount [8].

As shown in Fig. 2.1, landscape of power utilities encompasses a vast array of generation, transmission and distribution entities [9]. Continued research and development in protection methodologies are essential to address evolving challenges and ensure the resilience and reliability of power systems. These utilities are intricately interconnected, with distribution utilities linked to upstream transmission counterparts, which, in turn, draw power from generating utilities. Given this complex network, a standardized approach to designing protection schemes across these utilities is paramount for ensuring reliable operations. In India, established guidelines from esteemed organizations such as the IEEE [10], CIGRE and the Central Board of Irrigation and Power [11] serve as foundational frameworks for protection scheme design. However, the implementation of these guidelines presents challenges for power utilities, primarily due to the diverse range of relaying suppliers available in the market. Harmonizing diverse equipment and technologies to adhere to standardized protection protocols requires concerted effort and coordination among stakeholders.

Efforts to streamline the adoption and implementation of these guidelines are essential to enhance the reliability and effectiveness of protection schemes across the Indian power sector. Collaborative initiatives, knowledge sharing and leveraging advancements in technology can facilitate the seamless integration of standardized protection practices, thereby bolstering the resilience of India's power infrastructure.

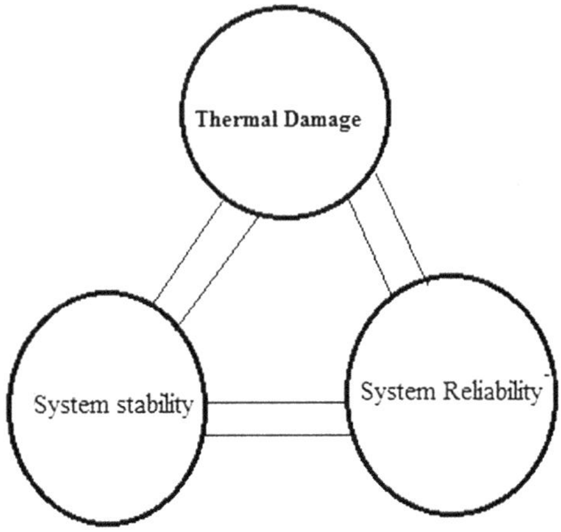


**FIGURE 2.1 Fundamental protection requirement.** It illustrates the triangular protection requirements discussed before, which serve as the foundation for planning and designing system protection in most power utilities.

## 2. Power system protection in practice

Power utilities are actively seeking optimal, fast, sensitive, selective, reliable and cost-effective protection solutions to safeguard their infrastructure. As depicted in Fig. 2.2, the functional block diagram illustrates the core components utilized in protection systems across power utilities. Both generating and transmission utilities aspire to implement two independent sets of protection [12,13]. This approach minimizes costs associated with additional coupling devices by employing independent protection cores. To mitigate expenses related to circuit breakers, independent tripping coils are utilized as a compromise, reducing the need for additional circuit breakers.

Merging units serve as replacements for traditional protection control wires, receiving fault signals from various protection cores. These signals are then aggregated by marshalling units, facilitating communication with relaying panels. Relays within these panels process fault information, with relay decisions transmitted to bay circuit breakers through a master trip relay. The master trip relay plays a crucial role in fault isolation between relaying panels and bay circuit breakers. In the event of a bay circuit breaker failure, redundancy is ensured through the presence of a station circuit breaker. Additionally, local end circuit breakers are backed by remote circuit breakers, enhancing overall system reliability and fault tolerance. This integrated approach to protection infrastructure underscores the commitment of power utilities to uphold operational integrity and minimize downtime.

The protection timing for protection requirement for different power utilities is shown in Fig. 2.3. The main protection is designed to clear the fault within 100 msec. Numerical relays are competent to compute a fault decision

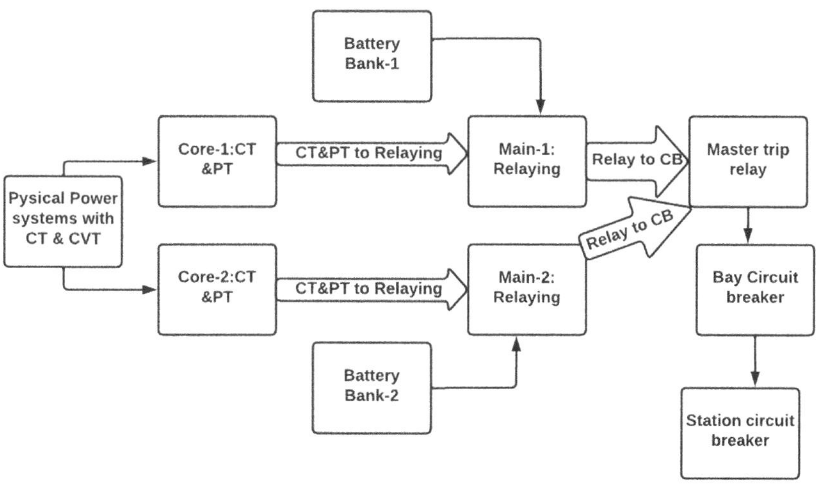


**FIGURE 2.2** Protection logic sequence.

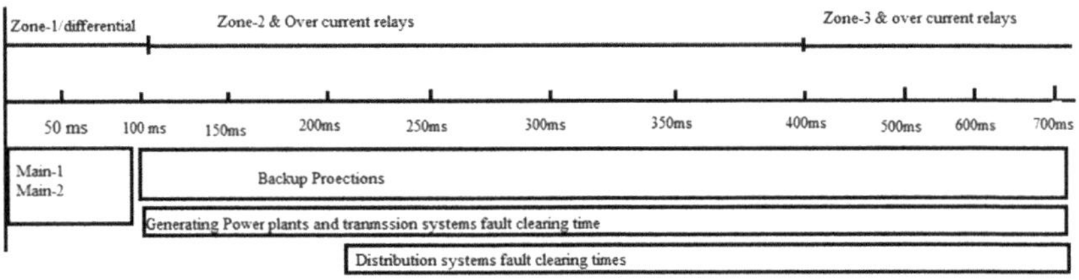


**FIGURE 2.3** Protection timing logic sequence.

within 20 msec, and a circuit breaker can clear a fault within 45msec. So, the minimum time required to clear the fault is around 100 msec. As discussed before, the generating units are protected with fast main-1 and main-2 protection relaying panels. These include unit differential and zone-1 distance relays. Generator transformers and transmission lines are also protected with two independent main protection designed to clear the fault with in 100 msec. Distance-based backup protection is provided for the protection of transmission lines through zone-2 and zone-3 operations. The fault on the distribution side is less severe, and system stability is not a big concern, so the low cost over current protection is generally used for fault clearance before the thermal damage of the equipment. In power utilities, protection practices are as classified as under.

## 2.1 Generating units

Synchronous machines are widely employed for power generation purposes [14], characterized by typical sub-transient direct axis reactance values ranging from 18%–20%. These machines are grounded through neutral grounding transformers [15], with a typical thermal time rating of around 2.0 s. Their continuous negative power sequence capability typically ranges from 8% to 10%. The reactive power feeding capabilities vary between 40%–60% of the active power rating, depending on the specific type of generating machine.

Given their critical role in power generation, synchronous machines require prompt protection to mitigate high fault stresses [16]. However, existing protection mechanisms often necessitate a minimum of 100 ms to clear faults. Protection measures for generating units encompass safeguarding against electrical faults in both stator and rotor windings, abnormal operations resulting from loss of excitation, over/under speeding, over/under frequency, unbalancing, overloading and de-rated operation [17].

The operational characteristics of synchronous machines are depicted in Fig. 2.4, illustrating their four-quadrant operation. In quadrants Q-1 and Q-IV, synchronous machines operate in forward generating mode, where they are protected against electrical stator faults, over/under frequency, unbalancing

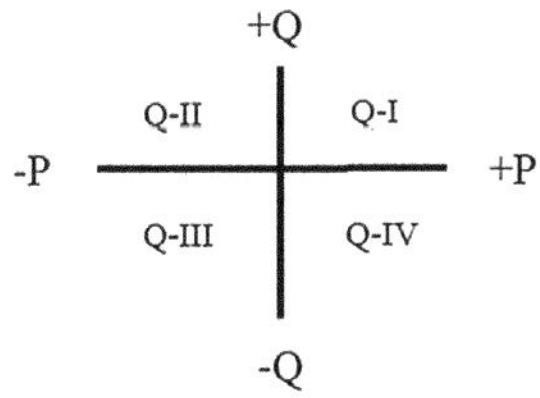


**FIGURE 2.4** Operation of synchronous machine.

and overloading. Conversely, in quadrants Q-III and Q-IV, these machines function as motors, requiring protection measures to promptly isolate them to prevent mechanical and electrical damage.

Synchronous machines are safeguarded through isolation from the main power systems by several means. This includes the opening of specific circuit breakers such as the extra-high circuit breaker (GT HV CB) located on the high-voltage side of the generator transformer, which isolates the generator from the main grid. Additionally, the field breaker (Field CB) is employed for excitation isolation, while the unit auxiliary transformer circuit breaker (UAT CB) and turbine input are also opened as part of the isolation process.

Generating units serve as the sole source of fault current in power plants or on power networks. However, isolating machines from the grid in the event of side faults is not always desirable. Hence, various classes of machine tripping are implemented to discern and discriminate the tripping of generating units [18]. These tripping classes enable selective and targeted protection, ensuring that only the necessary machines are isolated while maintaining the integrity and stability of the overall power system.

In Class-A tripping, immediate isolation of generating units from the main grid is essential in response to F3 faults, as illustrated in Fig. 2.5. This involves the swift opening of the GT HV CB, UAT CB, Field CB and turbine valve to prevent damage or excessive stress on the machine. By initiating these protective measures promptly, the integrity of the generating unit can be preserved, ensuring minimal disruption to operations and mitigating the risk of further damage to the system.

In Class-B tripping, if there's a mechanical fault (F2) detected in the turbine section of the generating unit, immediate action is taken to slow down the machine by operating the steam valve. Subsequently, the GT HV CB, UAT and Field CB are opened simultaneously to isolate the machine and prevent mechanical damage.

In Class-C tripping, generating units continue to operate at house load even in the event of an F1 grid-side fault, achieved by opening the GT HV CB breaker. The units supply power to their auxiliary loads through the UAT. After the network fault is cleared, the generating unit is ramped up to full rate without the need for re-synchronization. In some larger generating power plants, the use of a per-unit UAT is avoided to save costs. In such cases, Class-C tripping is achieved by opening a generator circuit breaker (GCB).

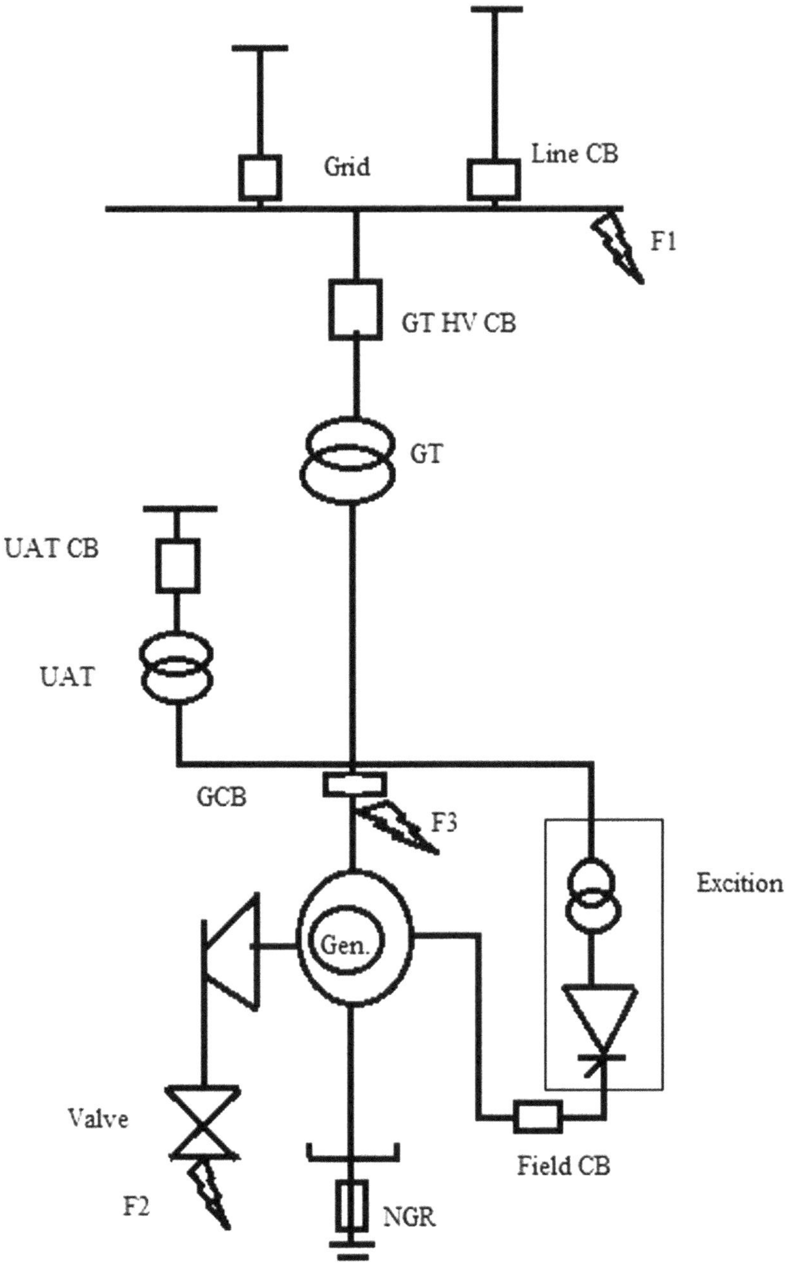


**FIGURE 2.5** Generator protection.

## 2.2 Generator protection philology

Protection of generating units involves various techniques to detect and respond to faults and abnormal conditions effectively. Stator earth and phase faults are common and require protection in all sizes of machines. Generator differential protection is commonly used for phase and earth fault protection, supplemented by high impedance earth fault protection in cases where matching current transformers (CTs) are unavailable. Additionally, zero sequence voltage/current detection aids in earth fault detection, especially for faults close to neutral points where fault signatures may be poor. Inter-turn faults are detected using either differential or zero sequence voltage/current methods depending on the winding type.

In contrast, the rotor, being ungrounded and free to rotate inside the stator, poses unique challenges. While the first fault in the field winding is typically insignificant, subsequent faults can lead to rotor field damage. Current-based protection, such as field circuit protection, is employed to detect unbalanced operations by computing the percentage of current sequence in terms of positive sequence. Over flux conditions are monitored through inducted voltage-to-frequency relations, with inverse relay mapping coordinated accordingly. Loss of field excitation is detected using impedance relays by monitoring changes in reactance trajectory.

Various abnormal operations, including over/under speeding, over/under voltage, fuse failure, dead machine and low forward/reverse power, are detected by measuring corresponding electrical parameters to activate the appropriate relay. Coordination among different protection functions is crucial to ensure effective and reliable machine protection, with fault trip logic employed to trip utilities operating within specified relay timeframes.

Popular protection relays for generating units include ABB REG670, ALSTOM P34X and SIEMENS 7UM6XX. Typically, two main relays, such as main-1 and main-2, are installed in generator relay panels 1 and 2, respectively, with protection functions grouped accordingly (see Table 2.1). Proper design and planning of generating unit protection require skilful coordination to enable fault detection and response efficiently, ensuring the continued safe and reliable operation of power generation systems.

## 2.3 Substation protection review

The lines, interconnecting transformers and reactors are major static power components at substations. The transformer [19,20] is a localized element with two sets of delicate winding as shown in Fig. 2.6. These winding may directly or indirectly experience the fault current. The transformer's main protection is designed to trip for transformer's internal fault and fault on the transformer bay [21]. For through fault, the transformer backup protection is designed to trip with a time delay. However, for other abnormal operations such as under/over-voltage and over-excitation, transformer protection is designed for alarm only.

**TABLE 2.1 Generator protection in Practice.**

| Group-1 | | Group-2 | |
|---|---|---|---|
| **Protection** | **ANSI** | **Protection** | **ANSI** |
| Generator differential | 87T | Over all differential | 87OT |
| Backup impedance | 21G | 95% stator earth fault | 61S1 |
| Negative phase Sequence | 46G | Loss of excitation | 40G |
| Over load | 51G | Out of step | 78G |
| 100% stator earth fault | 61S2 | Over/under-frequency | (81O/U) |
| Rotor earth fault | 64R | Reverse/low forward power | (32/37G) |
| Over/under-voltage | 59/27G | Over-fluxing | 24G |
| Inter-turn fault | 95G | Dead machine | 50DGM |
| – | – | Fuse failure(60) | 60 |

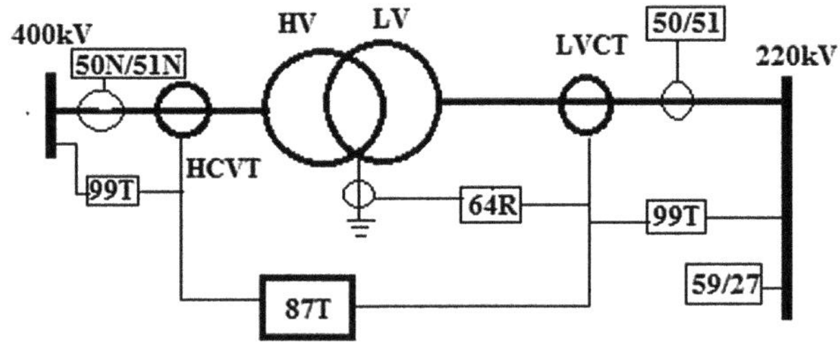

FIGURE 2.6 Diagram of transformer.

The power utilities are following the main-1 and main-2 protection philosophy for the protection of power transformers of class 220 kV and above voltage class as listed in Table 2.2. The main-1 protection will be differential protection for phase fault, over flux protection. Earth fault over current backup protection. The main-2 protection is enabled for restricted earth fault and backup over current protection for phase fault. This protection philosophy ensures the coverage of phase and earth fault from either main-1 or main-2 protection panels.

## 2.4 Transmission lines

Transmission lines are classified based on source to line impedance ratio (SIR). When SIR is less than 0.5, a line is classified as a long line. For lines

**TABLE 2.2 Transformer protection in practice.**

| Group-1 | | Group-2 | |
|---|---|---|---|
| **Protection** | **ANSI** | **Protection** | **ANSI** |
| Transformer differential | 87T | Over-current | 50/51 |
| Over-current backup | 50N/51N | Restricted earth fault | 64R |
| Over flux | 99T | Over/under-voltage | 59/27 |
| – | – | Thermal overload | 49T |

with SIR more than 4, the line is classified as a short line. For SIR 0.5–4, the line is classified as medium line.

The protection philology for transmission lines is based on the voltage class [22]. The fundamental requirement of transmission protection is to protect the main line against electrical faults occurring on the line and to provide backup protection to upstream transmission lines [23]. The coverage of line protection is restricted up to the primary side of the transformer installed at the remote end of the sub-station as shown in Fig. 2.6. Transmission lines of voltage 220 kV and above are protected by main 1 and main 2 protection, philology [24]. Both main 1 and main 2 are fast protection to clear the fault at 100 msec. These may include either the zone 1 distance relaying or carried based differential protection. Power utilities are utilizing the different types of main 1 and main 2 for better performance. It is observed that usually the main 1 protection is provided by-mho based MiCOM P44X series distance relays and the main 2 is provided by quadrilateral based Hitachi energy-make REL 670 or some other make of Siemen/GE make. Fig. 2.7 below gives a distance protection function enabled in distance relays for the line protection [25].

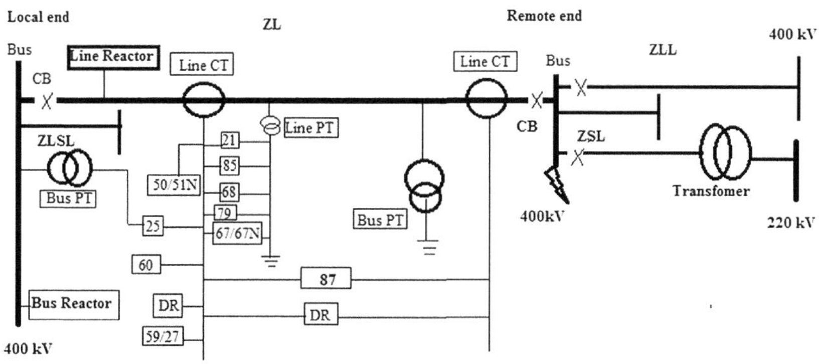


FIGURE 2.7 Line protection.

The protection scheme and zone setting philology for lines are tabulated in Tables 2.3–2.5 [26,27]. The ZL is the impedance of a protected line, ZLSL is the shortest line at the local end, ZSL is the shortest line at the remote end and ZLL is the impedance of the longest line at the remote end as shown in Fig. 2.7 [28].

Resistive reach for phase $(R_{ph})$ and earth $(R_g)$ are set to provide maximum coverage against all types of phase and earth faults with a check on load encroachment feature. These reaches are set in coordination with the respective resistive component of zone setting $(Z_{zone})$ and resistive load component [29] $(R_{load})$ of load impedance $(Z_{load})$. Tower footing resistance $(R_{tower})$ and arc resistance $(R_{arc})$ are also considered for reach setting calculations. As per NERC [30,31], Systems Protection and Control Committee, load impedance is calculated as Eqs. (2.1) and (2.2).

$$Z_{load} = 0.85 * V_{l\_l}/1.5 * I_{max} \tag{2.1}$$

$$Z_{(load)} = R_{load} + jX_{load} \tag{2.2}$$

$V_{l\_l}$ is line voltage and $I_{\max}$ is the maximum load current.

Line [32,33] and bus reactors [34] are also installed at substations for the regulation of system voltage within ±5% of the IEGC limit. These devices are single winging and have a high percentage impedance of order 120% of their name plate rating [35,36]. Their protection against the electrical faults is provided by differential (87), impedance protection (21) and over current earth fault (50N/51N).

Differential protection [37] is used for the protection of bus bars. The reverse zone of line distance relays is also set to provide the backup for the bus bar faults. The typical setting of the bus bar protection is maximum load on

**TABLE 2.3 Line protection in practice.**

| Group 1 | | Group 2 | |
|---|---|---|---|
| **Protection** | **ANSI** | **Protection** | **ANSI** |
| Distance relay | 21 | Distance relay | 50/51 |
| Carrier pilot communications | 85 | Instantaneous/time-delay earth current relay | 50N/51N |
| Instantaneous/time delay over current relay | 50/51 | AC dDirectional overcurrent relay | 67 |
| Differential | 87L | AC reclosing relay | 79 |
| Power swing blocking | 68 | Over/under-voltage | 59/27 |
| Synchronizing | 25 | – | – |

**TABLE 2.4 Zone setting in practice.**

| Zone | Zone setting | Time (sec) |
|---|---|---|
| Zone 1 | Z1 = 80%*ZL | 0.1 |
| Zone 2 | Z2 = 120%*ZL(SC) | 0.5 |
| | Z2 = 150%*ZL(DC) | 0.35 |
| Zone 3 | Z3 = 120% *(ZL + ZLL) | 1.0 |
| Zone 4 | Z4 = 50%*ZLSL | 0.5 |
| Power swing | Block zones except zone I, block zone and trip with out of step (OOS) function | 2.0 |
| Inter tripping | Carried aided permissive under/over-reach | 0.1 |
| Broken conductor | 20% of ibase. | 20 |
| Auto re-closing | Single pole trip and reclosing dead time | 1.0 |
| Dead time | Reclaim time | 25 |
| LBB | Current setting I > 20% | |
| Back-up protection | Over-current protection | 1.0 |
| | 220 kV and lower-voltage level IDMT O/C and E/F | |

**TABLE 2.5 Resistive reach settings.**

| Make | Resistive phase reach ($R_{ph}$) | Resistive earth reach ($R_g$) |
|---|---|---|
| MICO M | $R_{ph} = \min(0.6 * R_{load}, 10 * R_{zone})$ | $R_g = \min(0.8 * R_{load}, 10 * R_{zone})$ |
| Siemen | $R_{ph} = (R + R_{arc})_{zone}\ 2$ | $R_g = (R_{zone} + R_{tower} + R_{arc})$ |
| Hitachi energy | $R_g = (3 * R_{zone}, 2 * R_{arc})$ | $R_g = (4.5 * R_{zone}, R_{tower} + R_{arc})$ |

anyone connected to the bus bay [38]. The reverse zone 1 is set to cover 50% of the shortest line connected at the local end (ZLSL).

Other protection infrastructures installed for protection are independent dual set of battery bank, synchronization facility, PLCC, event logger, centralized data recorder, etc.

## 3. Relay setting scheme and protection setting for a generating plant: Case study

Design and review of setting calculations for generating power, which consists of generating units, generator transformers, bus bar and transmission lines evacuating power to the grid, is an ideal choice for planning the protection. A generating power unit is 100 MW. The other details of the generating unit, generator transformer and a transmission line are listed in Table 2.6.

The main protection functions and setting are derived as per regional power committee and CBIP guidelines as listed in Table 2.7.

## 4. Future of protection

The existing numerical relays are very reliable, secure and selective for sensing the power systems' short-circuit faults. Deployment of main 1 and main 2 protection approaches has eliminated any change of protection systems failure.

The foundation of prevailing protection is the huge contribution of short circuits from synchronous machines feeding power to power systems [6].

**TABLE 2.6** System parameters.

| Generator | Apparent power | 111.1 | MVA |
|---|---|---|---|
| | Active power | 100 | MW |
| | Terminal voltage | 13.8 | kV |
| | Transient (Xd′) | 0.25 | P.U |
| | Sub-transient (Xd″) | 0.2 | P.U |
| | Phase CT ratio | 6000/5 | – |
| | Neutral CT ratio | 6000/5 | – |
| | PT ratio | 13.8/110 | kV/V |
| Transformer | Apparent power | 420 | MVA |
| | Voltage rating | 13.8/400 | kV |
| | Impedance | 14.5 | P.U |
| | LV ratio | 6000/1 | – |
| | HV ratio | 600/1 | – |
| Protected transmission | Length | 2.74 | kM |
| | Positive seq. | 0.0298 | Ω/p |

**TABLE 2.7 Summary of protection setting.**

| Element | Protection | Protection function | Setting | Time |
|---|---|---|---|---|
| | | Pick-up | | |
| | | Low | 1A | |
| | | Setting | | |
| | | Pick-up | | |
| | Generator | High | 5A | |
| | Differential | Setting | | 0.1 |
| | Protection | Slope 1 | 25% | |
| | | Slope 2 | 50% | |
| | | Base | | |
| | | Point of | 2.5 | |
| | | Slope 1 | | |
| Generator | Generator | Pick up | 0.50 | |

Increasing deployment of inverter interfaced renewables [6,39] in power systems will have a significant impact on the systems' short-circuit strength [40]. The dominance of renewable generation may be quite high, and the performance of the prevailing numbers of relays may deteriorate.

The expected fault signatures for futuristic power systems will be low fault levels, deteriorating power quality, transient [40], etc. Adaptive relaying techniques which can adopt the relay setting as per changing fault level may be one of the promising solutions [41—43]. A dynamic state estimation based protection may also be an alternative protection solution [43]. The adaptive relays with multiple group setting have gained application in industries for the protection of wind in feed farms with intermittent and time variant fault in feed. These schemes work on the identification of dominant network topologies and updating of coordinated relay group setting for the desired protection coordination. In solar power farms, the group-setting-based schemes and differential relays are gaining the popularity. The voltage—current time inverse relaying schemes are also has potential for the protection of solar farms.

## 5. Conclusion

The planning of protection is important for the design of a reliable power system. Coordination of system protection with upstream generating systems and downstream distribution systems is crucial. This chapter discusses the prevailing protection practice among the power generating and transmission utilities. Power utilities are following dual main protection philology for 220 kV and the above system voltage. A single main and backup protection approach is practiced for network voltage below 220 kV.

A few relay setting parameters are vender specific, and power utilities are facing a great problem in their setting due to non-availability of uniform setting guidelines. This chapter compiles a summary of relay setting of these relays for the ease of utilities. The emerging challenges and futuristic research in area of low fault feed resources are also briefed in this chapter.

## References

[1] O.L. Elgard (Ed.), Electrical Energy Systems Theory-An Introduction, second ed., 2009.

[2] S. Manohar, T. Vishnuvardhan, S.G. Srivani, Adaptive protection coordination scheme for power networks under penetration of distributed energy resources, IET Gener.Trans.Distrib. 10 (15) (2016) 3919–3929.

[3] M. Ezzeddine, R. Kaczmarek, M.U. Iftikhar, Coordination of directional overcurrent relays using a novel method to select their settings, IET Gener. Transm. Distrib. 5 (7) (2011) 743–750.

[4] M. Singh, V. Telukunta, S.G. Srivani, Enhanced real time coordination of distance and user defined over current relays, Int. J. Electr. Power Energy Syst. 98 (2018) 430–441.

[5] IEEE Guide for AC Generator Protection, ANSI/IEEE C37, 2006, p. 102.

[6] Y. Fang, K. Jia, Z. Yang, Y. Li, T. Bi, Impact of inverter-interfaced renewable energy generators on distance protection and an improved scheme, IEEE Trans. Ind. Electron. 66 (9) (Sep. 2019) 7078–7088.

[7] A. Banaiemoqadam, A. Hooshyar, M.A. Azzouz, A control-based solution for distance protection of lines connected to converter-interfaced sources during asymmetrical faults, IEEE Trans. Power Deliv. 35 (20) (June 2020) 1455–1466.

[8] IEEE Power System Relaying Committee WG D6, Power Swing and Out-of-Step Considerations on Transmission Lines, July 2005. http://www.pes-psrc.org.

[9] J. Pradhan, M. Singh, Modified distance relaying blocking scheme for low inertia system, in: 2020 21st National Power Systems Conference (NPSC), Gandhinagar, India, 2020, pp. 1–6.

[10] IEEE Tutorial on the Protection of Synchronous Generators, second ed., Special Publication of the IEEE Power System Relaying Committee, 2010.

[11] Manual on Power System Protection, Central Board of Irrigation and Power, January 2018.

[12] IEEE Recommended Practice for Grounding of Industrial and Commercial Power Systems, IEEE Std, 1991, p. 142.

[13] Protection Considerations for Combustion Gas Turbine Static Starting; Working group J-2 of the rotating machinery subcommittee, power system relay committee.

[14] Protective Relaying for Power Generation Systems; Donald Reimert, CRC Press, 2006.

[15] Practical improvement to stator ground fault protection using negative sequence current, in: R. Patterson, A. Eltom (Eds.), IEEE Transactions Paper Presented at the Power and Energy Society General Meeting (PES), IEEE, 2013.

[16] Behavior Analysis of the Stator Ground Fault (64G) Protection Scheme, R. Sandoval, F. Morales, E. Reyes, S. Meléndez, J. Félix, Presented to the Rotating Machinery Subcommittee of the IEEE Power System Relaying Committee, January 2013.

[17] Advanced Generator Ground Fault Protections, W. Hartmann, Presented at the Western Protective Relay Conference, October 2015.
[18] IEEE standard for electrical power system device function numbers, acronyms, and contact designations, IEEE Standard C37 (2008) 2.
[19] Z.S. Almajali, A discrimination method between transformer inrush and fault current based on chromatic monitoring, WSEAS Trans. Power Syst. 16 (2021) 308.
[20] IEEE guide for protecting power transformers, in: IEEE Std C37.91-2021 (Revision of IEEE Std C37.91-2008), June 29, 2021, pp. 1–160.
[21] J. Ciufo, A. Cooperberg, Transformer protection, in: Power System Protection: Fundamentals and Applications, IEEE, 2022, pp. 185–235.
[22] S. Zubic, Z. Gajic, D. Kralj, Line protection operate time: how fast shall it be, IEEE Access 9 (2021) 75608–75616.
[23] N.O. Mogaru, The basics of transmission line protection, in: 2020 IEEE PES/IAS Power Africa, Nairobi, Kenya, 2020, pp. 1–5.
[24] R. Moxley, K. Fodero, H.J. Altuve, Updated transmissionlineprotectioncommunications, in: 2009 62nd Annual Conference for Protective Relay Engineers, College Station, TX, USA, 2009, pp. 394–401.
[25] G. Ziegler, Numerical Distance Protection: Principles and Applications, Wiley, 2011.
[26] Technical standards for connectivity to the grid central electricity authority regulation feb 2007.
[27] D.K. Ibrahim, G.M. Abo-Hamad, E.E.-D.M.A. Zahab, A.F. Zobaa, Comprehensive analysis of the impact of the TCSC on distance relays in interconnected transmission networks, IEEE Access 8 (2020).
[28] K. Jia, T. Bi, Y. Fang, W. Li, Q. Yang, Ground fault distance protection for paralleled transmission lines, IEEE Trans. Ind. Appl. 51 (6) (November/December 2015) 5228–5236.
[29] B.R. Bhalja, R.P. Maheshwari, High-resistance faults on two terminal parallel transmission line: analysis, simulation studies, and an adaptive distance relaying scheme, IEEE Trans. Power Deliv. 22 (2) (April 2007) 801–812.
[30] D.A. Tziouvaras, S.R. Chano, J.-L. Chanelière, J.H.S. Moroni, M.S. Sachdev, et al., Modern Distance Protection Functions and Applications Paris France, CIGRE Publications, October 2008.
[31] B. Taheri, M. Sedighizadeh, Detection of power swing and prevention of mal-operation of distance relay using compressed sensing theory, IET Gener. Transm. Distrib. 14 (23) (December 2020) 5558–5570.
[32] M.M. Eladany, A.A. Eldesouky, A.A. Sallam, Power system transient stability: an algorithm for assessment and enhancement based on catastrophe theory and FACTS devices, IEEE Access 6 (2018) 26424–26437.
[33] S.M. Mohan, S. Chatterjee, Busbar protection—a review, in: 2010 IEEE Region 8 International Conference on Computational Technologies in Electrical and Electronics Engineering (SIBIRCON), Irkutsk, Russia, 2010, pp. 755–759.
[34] Z. Gajić, H. Faramawy, L. He, K. Koppari, L. Max, M. Kockott, Modern design principles for numerical busbar differential protection, in: 2019 72nd Conference for Protective Relay Engineers (CPRE), College Station, TX, USA, 2019, pp. 1–16, https://doi.org/10.1109/CPRE.2019.8765881.
[35] D. Mourad, E.H. Shehab-Eldin, Simple and adaptive busbar protection scheme considering CT saturation effect, in: 2017 Nineteenth International Middle East Power Systems Conference (MEPCON), Cairo, Egypt, 2017, pp. 71–77.
[36] IEEE Guide for Breaker Failure Protection of Power Circuit Breakers, July 2016, pp. 1–73.

[37] IEEE guide for the protection of shunt reactors, in: IEEE Std C37.109-2006 (Revision of IEEE Std C37.109-1988), April 23, 2007, pp. 1−39.

[38] V. Telukunta, J. Pradhan, A.M. Singh, S.G. Srivani, Protection challenges under bulk penetration of renewable energy resources in power systems: a review, CSEE J. Power Energy Sys. 3 (4) (December 2017) 365−379.

[39] P.K. Nayak, A.K. Pradhan, P. Bajpai, A three-terminal line protection newline scheme immune to power swing, IEEE Trans. Power Deliv. 31 (3) (June 2016) 999−1006.

[40] B.A.M. Alizadeh, M. Khederzadeh, R. Razzaghi, Fault detection during power swing in thyristor- controlled series capacitor-compensated transmission lines, Elec. Power Syst. Res. 187 (October 2020).

[41] B. Sahoo, S.R. Samantaray, System integrity protection scheme for enhancing backup protection of transmission lines, IEEE Syst. J. 15 (3) (August 2020) 4578−4588.

[42] J. Zhao, et al., Power system dynamic state estimation: motivations, definitions, methodologies, and future work, IEEE Trans. Power Syst. 34 (4) (July 2019) 3188−3198.

[43] M.T. Bishop, S.R. Mendis, J.F. Witte, K.L. Leix, Selecting overcurrent protection for three-phase transformers, IEEE Ind. Appl. Mag. 2 (2) (March-April 1996) 35−41.

## Further reading

[1] T. Hoang, Q.T. Tran, Y. Besanger, An advanced protection scheme for medium-voltage distribution networks containing low-voltage microgrid with high penetration of photovoltaic systems, Int. J. Electr. Power Energy Syst. 139 (2022) 107988.

[2] T.S. Sidhu, H.S. Gill, M.S. Sachdev, A power transformer protection technique with stability during current transformer saturation and ratio-mismatch conditions, IEEE Trans. Power Deliv. 14 (3) (July 1999) 798−804.

Chapter 3

# DC type circuit breakers

**Vassilis C. Nikolaidis**
*School of Electrical and Computer Engineering, National Technical University of Athens, Athens, Greece*

## 1. What is a DC circuit breaker?

A definition for circuit breakers is [1]: 'A circuit breaker is a device capable of making, carrying and breaking currents under normal circuit conditions and also making, carrying for a specified time, and breaking currents under specified abnormal circuit conditions'.

According to Ref. [2], a DC circuit breaker is a device that is 'used to close and interrupt a DC power circuit under normal conditions, or to interrupt this circuit under fault or emergency conditions'.

One can generalize these definitions to specify that the main objective of any DC circuit breaker is to interrupt or eliminate overcurrents flowing through the protected DC circuit, which otherwise might cause permanent damage to the circuit equipment. During non-fault conditions, DC circuit breakers do not intentionally intervene in DC circuit's normal operation.

DC circuit breakers may be relatively simple devices, which base their operation on physics laws (electromagnetic forces, spring forces, etc.), or may become very complex systems, especially when they comprise power semiconductors. For instance, mechanical circuit breakers (MCBs) are commonly used in low-voltage DC (LVDC) applications. MCBs are self-contained devices forming the simplest (but not simple) type of a DC circuit breaker. In high-voltage DC (HVDC) applications, DC circuit breakers are much more complicated systems consisting of many different components (power electronic devices, cooling system, driving system, controllers, surge arrester).

In most cases, the operation of DC circuit breakers relies on current based quantities. Typical actuating quantities are the current magnitude, the rate of change of current magnitude, or both. There are also other quantities that are used to detect and break fault currents, such as voltage measurement or impedance calculation.

**Switchgear Design, Operation, and Maintenance Using Industry Standards**
**https://doi.org/10.1016/B978-0-443-24718-7.00003-2**

In principle, a DC circuit breaker has the structure of Fig. 3.1, consisting of three distinct operational branches:

- The continuous current branch.
- The commutation current branch.
- The energy dissipation branch.

Under normal circuit operation conditions, current flows through the continuous current branch of the DC circuit breaker. When abnormal circuit conditions are detected, the current is first directed to pass through the commutation current branch and afterwards through the energy dissipation branch to dissipate the residual magnetic energy stored in the system.

In some types of DC circuit breakers, the commutation current branch may not be physically present (e.g., in solid-state circuit breakers). In other types, there is no clear spatial distinction between the three branches (e.g., in low-voltage MCBs). A more detailed description of the DC circuit breakers will be given in Section 4.

## 2. DC fault current interruption

To analyze DC current interruption, let us consider the simple circuit of Fig. 3.2. In this circuit, a DC voltage source supplies a load through a line with equivalent resistance $R$ and inductance $L$. A DC circuit breaker (shown with the blue dashed line) is connected in series with the line to interrupt the current if a short-circuit fault occurs.

Let us consider a bolted short-circuit fault at load terminals, as shown in Fig. 3.2. Applying Kirchhoff's voltage law, the following equation derives:

$$U_{DC} - Ri(t) - L\frac{di(t)}{dt} - U_{CB} = 0 \tag{3.1}$$

where $i(t)$ is the short-circuit current and $U_{CB}$ is the voltage across the circuit breaker.

Voltage $U_{CB}$ is almost zero when the contacts of the mechanical switch (MS) are closed. After contacts separation, considerable voltage $U_{CB}$ is being

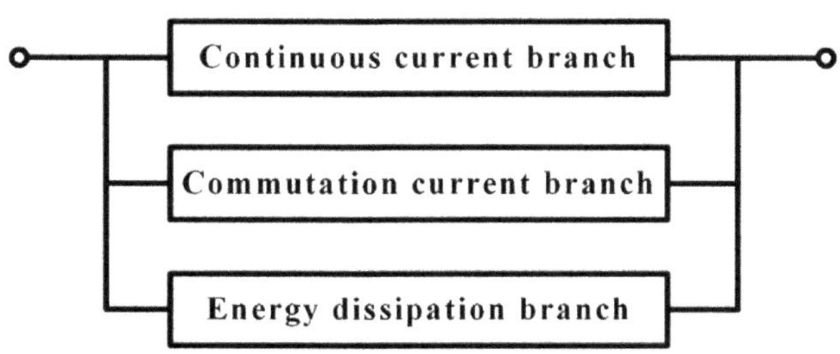


**FIGURE 3.1 Structure of a DC circuit breaker.** Functional description of a DC circuit breaker.

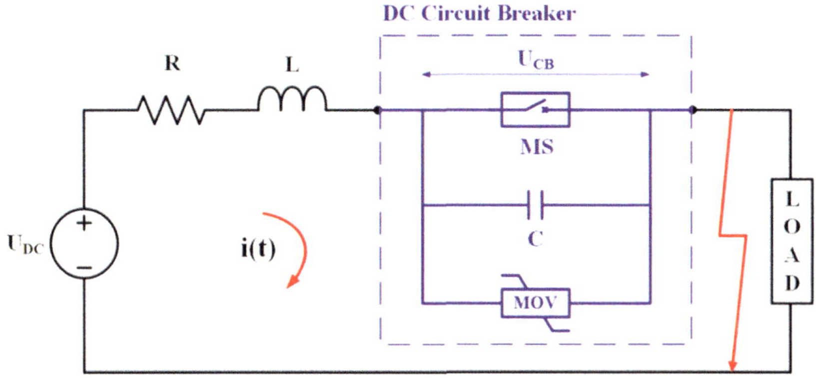


**FIGURE 3.2 DC circuit breaker in a faulty circuit.** A short-circuit fault in a simple DC circuit with a DC circuit breaker.

developed across the contacts due to the system trying to release the magnetic energy stored in the equivalent inductance $L$. The stored magnetic energy is equal to

$$E = \frac{1}{2} L I^2 \tag{3.2}$$

where $I$ is the current flowing in the circuit just before the contacts open.

Neglecting resistance, the magnetic energy is transformed into electric energy by charging the capacitance $C$ (a physical capacitor and/or stray capacitance) that appears in parallel to the MS:

$$\frac{1}{2} C U_{CB}^2 = \frac{1}{2} L I^2 \tag{3.3}$$

Voltage $U_{CB}$ has an opposite polarity with respect to supply voltage $U_{DC}$. In circuit breaker theory, counter-voltage $U_{CB}$ is better known as transient interrupting voltage (TIV). This voltage is the main cause of any DC current interruption process.

In principle, TIV must be generated and maintained across the circuit breaker for a sufficient time. At the same time, it must be higher than the supply voltage $U_{DC}$ and lower than the voltage stress withstand capability of the capacitance. A metal oxide varistor (MOV) is usually connected in parallel to the MS to limit and control the voltage stress across the contacts.

Fig. 3.3 shows the response of voltages and currents during the short-circuit fault [3]. The circuit current $i$(t) increases after the inception of the fault, at time instant $t_0$. Based on (3.1), the rate of change of short-circuit current is given by

$$\frac{di(t)}{dt} = \frac{1}{L}(U_{DC} - Ri(t) - U_{CB}) \tag{3.4}$$

During the period from fault inception (at time instant $t_0$) until the TIV equals the supply voltage $U_{DC}$(at time instant $t_2$), voltage $U_{CB}$ can be considered equal to zero ($U_{CB} \approx 0$). By integrating (3.4), the following expression of the current flowing in the circuit during this period is obtained:

$$i_{t02}(t) = \frac{U_{DC}}{R}\left(1 - e^{-\frac{t}{\tau}}\right) + Ie^{-\frac{t}{\tau}} \tag{3.5}$$

where $\tau = L/R$ is the circuit time constant.

By assuming a constant voltage $U_{CB}$ equal to $U_{CLAMP}$ after time instant $t_2$, the current takes the following expression [4]:

$$i_{t23}(t) = \frac{U_{DC} - U_{CB}}{R}\left(1 - e^{-\frac{t}{\tau}}\right) + I_{\max}e^{-\frac{t}{\tau}} \tag{3.6}$$

where $I_{\max}$ is the maximum fault current, which progressively reduces to zero until time instant $t_3$. If no interruption occurred, the prospective fault current would have increased above $I_{\max}$ as can be seen with the dashed line in Fig. 3.3.

The assumption of a constant voltage across the breaker after $t_2$ is sufficient since the MOV clamps the voltage to the almost constant value $U_{CLAMP}$.

It can be seen from (3.6) that current reduction occurs only when $U_{CB}$ is greater than $U_{DC}$.

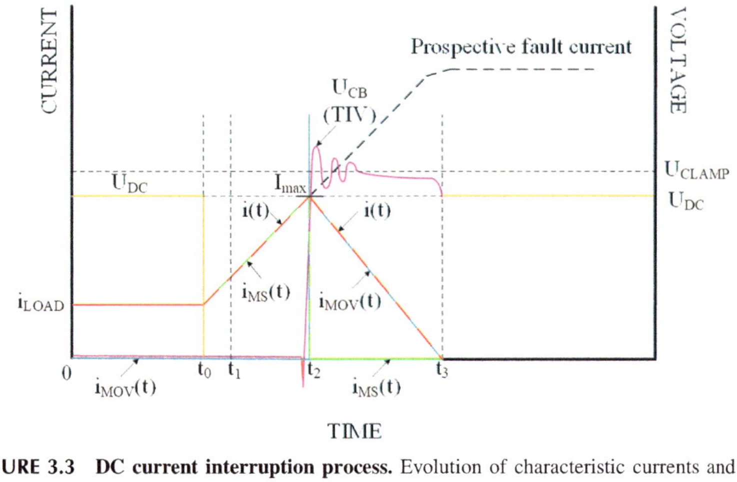


**FIGURE 3.3 DC current interruption process.** Evolution of characteristic currents and voltages in a DC breaker during a fault.

The total energy absorbed by the DC circuit breaker is given by

$$W = \frac{1}{2} L I_{\max}^2 \left(1 + \frac{U_{DC}}{U_{CB} - U_{DC}}\right) \quad (3.7)$$

This energy must be dissipated from the circuit breaker. In MCBs, the produced arc voltage makes sufficient TIV. In these cases, the stored magnetic energy is dissipated as arc thermal energy.

## 3. Requirements of DC circuit breakers

DC circuit breakers must meet three major technical objectives:

- During normal operation (i.e., during continuous, normal load current flow), DC circuit breakers must produce minimum losses.
  It is crucial to design DC circuit breakers with a very high efficiency. To minimize power losses, MSs are preferred in the continuous current branch. MSs comprise low resistance metallic contacts, which offer the most effective conduction path among all the switching technologies. On the other hand, if power electronic switches are used in the continuous current branch instead of MSs, power losses increase. This is due to the switching of the power semiconductor devices, which causes high on-state power losses. A compromise between power losses and interruption speed is made when MSs and power semiconductors are together used in the DC breaker.
- During abnormal operation (e.g., during short-circuit faults), DC circuit breakers must break the current flow as faster as possible.
  Due to low reactance of DC power systems, short-circuit fault currents increase very fast, and therefore, they must be interrupted in a few milliseconds. The rate-of-change of fault current ($d\,i/d\,t$) and the time instant ($t_{\text{peak}}$) the prospective fault current reaches its maximum value are two critical quantities that specify the required interruption speed of a DC circuit breaker. Both quantities greatly depend on the DC source characteristics. Fig. 3.4 illustrates typical short-circuit current responses for different types of DC sources: (1) capacitor with different RLC circuit characteristics, (2) battery, (3) rectifier-interfaced rotating ac source (generator, motor) [5]. When multiple sources are connected to a DC power system, the total short-circuit current can be taken by the superimposition of the short-circuit currents from all the sources. In principle, DC circuit breakers using MSs to interrupt overcurrents are slower than that using power semiconductor switches.
- During current interruption, DC circuit breakers must have the capability to dissipate the absorbed thermal energy.
  The produced thermal energy increases very quickly and proportionally to the square of the fault current. Delayed or insufficient thermal dissipation

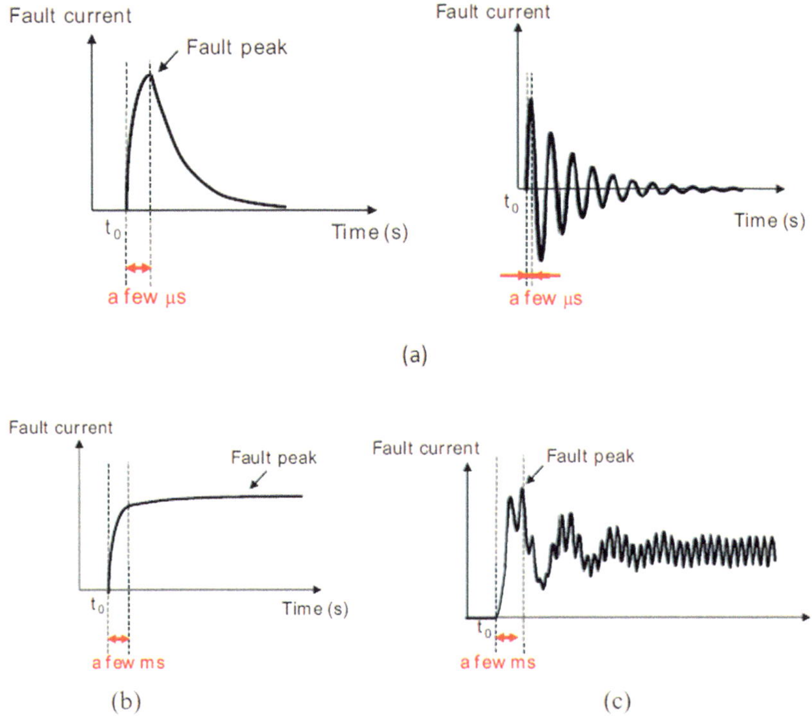


**FIGURE 3.4 Short-circuit current contribution of various DC sources: (a) Capacitor with $R > 2\sqrt{L/C}$, (left), Capacitor with $R \leq 2\sqrt{L/C}$ (right), (b) Battery, (c) AC rotating machine.** Typical short-circuit fault current waveforms for DC circuits with different elements.

may cause thermal damage of the DC circuit breaker. Typically, MOVs are connected in parallel to protect the circuit breaker. A MOV is a non-linear resistor whose resistance remains extremely high until the voltage across the device exceeds a specific value. Above this value, the resistance drops significantly forming a low resistance path for the current. During conduction, the voltage across the MOV remains almost constant at a value characterized as $U_{CLAMP}$.

Besides the aforementioned technical objectives, there are also several other requirements for DC circuit breakers. Among them, the most important is the capital and maintenance cost of a DC circuit breaker, which must be kept low, and its expected lifetime, which should be adequately long.

## 4. Classification of DC circuit breakers

DC circuit breakers are classified into the following categories:

- Mechanical circuit breakers (MCBs)
- Solid-state circuit breakers (SSCBs)
- Hybrid circuit breakers (HCBs)
- Inherent current limiters

## 4.1 Mechanical circuit breakers

MCBs include an MS, which opens its metal contacts when a fault is detected. Physically, the contacts are separated with each other by electromagnetic or thermal–magnetic forces produced by the overcurrent. During the opening, an arc is generated between the contacts. The respective arc voltage counteracts the supply voltage and helps reduce the fault current. If needed, more than one MS (i.e., electrical contacts) are connected successively to increase the generated arc voltage.

Very critical is how the arc is manipulated [6]. In air arc chute circuit breakers, the arc is forced into the air chute, where it is split to smaller arcs and is cooled to dissipate the generated thermal energy. After some time, the arc extinguishes leaving an open circuit. In air blast circuit breakers, compressed air is blown through the arc to cool it. In vacuum circuit breakers, the high dielectric strength of the vacuum gap provides proper conditions for sufficient arc quenching. $SF_6$ shows even better dielectric strength properties. $SF_6$ is a compressed gas, which is sprayed to the arc to cool it and decompose its by-products.

Fig. 3.5 illustrates the time response of voltage and current magnitudes during a short-circuit test on an MCB [7]. It can be seen that immediately after short-circuit inception, at time instant $t_o$, the fault current (shown with the

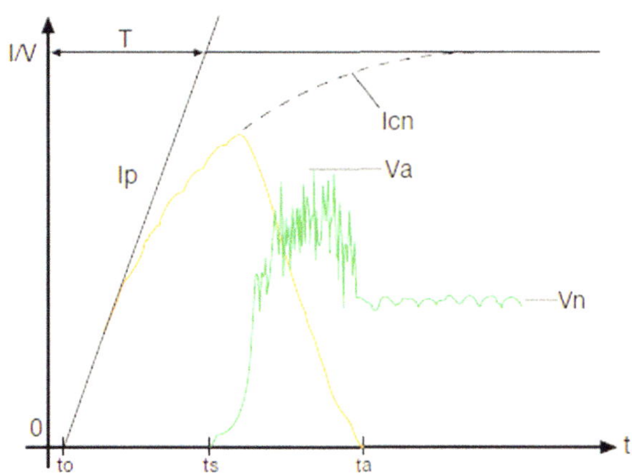


**FIGURE 3.5 Interruption process in an actual MCB.** Fault interruption in a real DC mechanical circuit breaker.

orange curve) rapidly increases. After time instant $t_s$, when the switch contacts open, arc voltage $V_a$ (shown with the green curve) is generated, counteracting the supply voltage $V_n$. Very shortly, after $V_a$ becomes greater than $V_n$, the fault current starts to reduce until it is extinguished, at time instant $t_a$. The prospective short-circuit current $I_{cn}$ is also shown with the dashed line in Fig. 3.5.

MCBs offer the lowest power losses among all the DC circuit breaker types. This is due to the fact that the metal contacts have low resistance, which ranges from a few μΩ to a few mΩ. Since MCBs consist of moving parts (contacts, springs, actuators) and are exposed to arcs, maintenance is more expensive, whereas their lifetime is limited by the maximum permissible number of breaking operations. For gas circuit breakers, there are also concerns for gas emission. Their size and weight are also critical for some MVDC applications [8]. The most critical disadvantage of MCBs is the response time, which is in some cases is considered too slow for DC current interruption. In fact, MCBs require tens to hundreds of milliseconds to interrupt a DC fault current.

MCBs are commonly used in LVDC applications, that is, in LV microgrids, electric vehicles, renewable energy plants, and buildings. They are also used in MVDC railway and traction systems.

## 4.2 Solid-state circuit breakers

SSCBs comprise power semiconductors in the continuous current branch (Fig. 3.6). The semiconductors are normally operating in on-state to permit the current to flow continuously through the DC circuit breaker under normal circuit conditions. When a short-circuit fault occurs, a voltage or current signal is submitted by the gate driver unit to the gate of the power semiconductors to turn them in the off-state. After this sudden change, voltage across the breaker increases rapidly due to the electromagnetic energy that is trapped in system's equivalent inductance. To protect the circuit breaker from this potentially dangerous transient overvoltage, an MOV is installed in parallel forming the energy dissipation branch. The MOV absorbs the energy until the current is

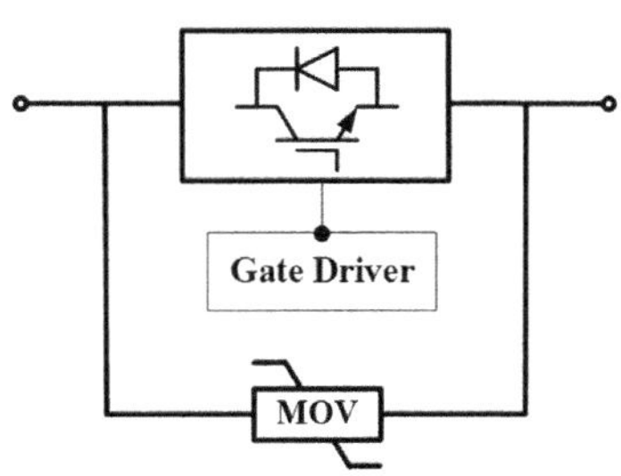


**FIGURE 3.6 SSCB schematic.** Generic structure of a DC solid-state circuit breaker.

driven to zero. Note that a distinct commutation current branch is not present in SSCBs.

SSCBs may be unidirectional or bidirectional [6]. Unidirectional SSCBs allow normal current flows in both directions but have the capability to interrupt fault currents flowing only to one direction. Fig. 3.7 illustrates a typical unidirectional SSCB. Bidirectional SSCBs allow and break currents flowing in both directions. If bidirectional current interruption is desired, then similar semiconductors group with an anti-parallel diode in anti-series must be used. Fig. 3.8 illustrates some typical configurations of bidirectional SSCBs. If high total voltage withstand capability is required (for instance, in MVDC applications), multiple semiconductor groups are connected in series, whereas also diodes are connected in series to increase the reverse block voltage of the device.

An alternative classification of SSCBs is based on their capability to operate in two or multiple modes: bi-mode or tri-mode SSCBs [5]. The bi-mode SSCBs are operating in two distinct modes: in on-state, during normal current flows, and in off-state, during faults. Besides these two typical operation states, tri-mode SSCBs also allow the overloaded operation of the converters for a specific short period of time. This allows the SSCBs to ride-through expected overcurrent conditions such as inrush currents.

A special type of SSCB is the so-called Z-source breaker. The Z-source breaker combines a thyristor with a Z-source circuit. There are several alternative topologies encountered for the Z-source breaker architecture. The most typical configuration is shown in Fig. 3.9A [9]. This topology experiences two different behaviours. During normal system operation, system's response is driven by the low-frequency equivalent circuit of Fig. 3.9B. On the contrary, the high-frequency circuit of c drives the response of the circuit breaker during faults. Due to the zero current in the inductor, the transient current in the capacitor drives the thyristor current to zero, meaning that it is switched off.

In general, SSCBs have the advantage of rapid fault clearing; fault currents are interrupted in a few μs. Moreover, arcs are avoided during fault clearing.

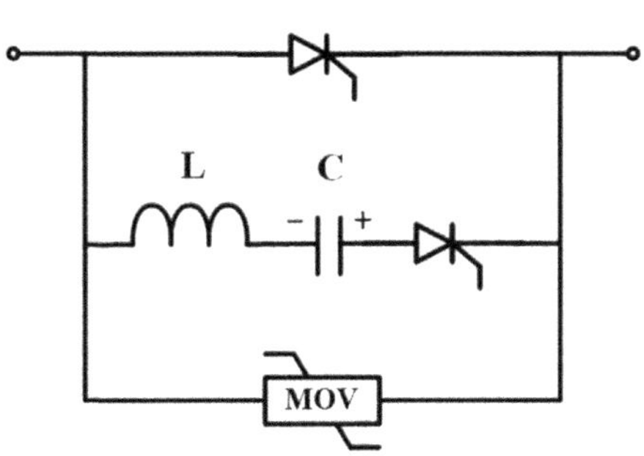


**FIGURE 3.7 Unidirectional SSCB.** Typical topology of a unidirectional DC solid-state circuit breaker.

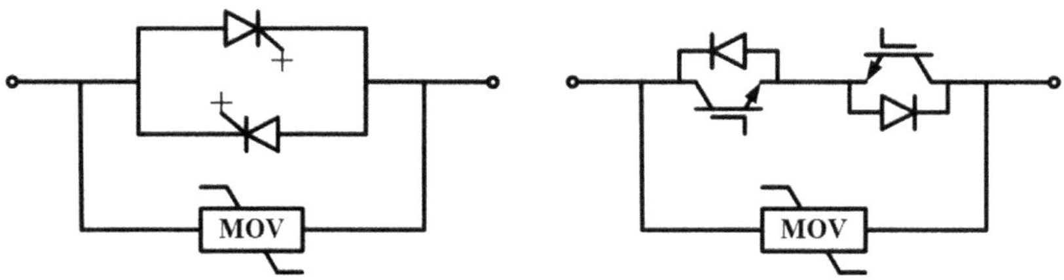


**FIGURE 3.8 Bidirectional SSCBs.** Typical topology of a bidirectional DC solid-state circuit breaker.

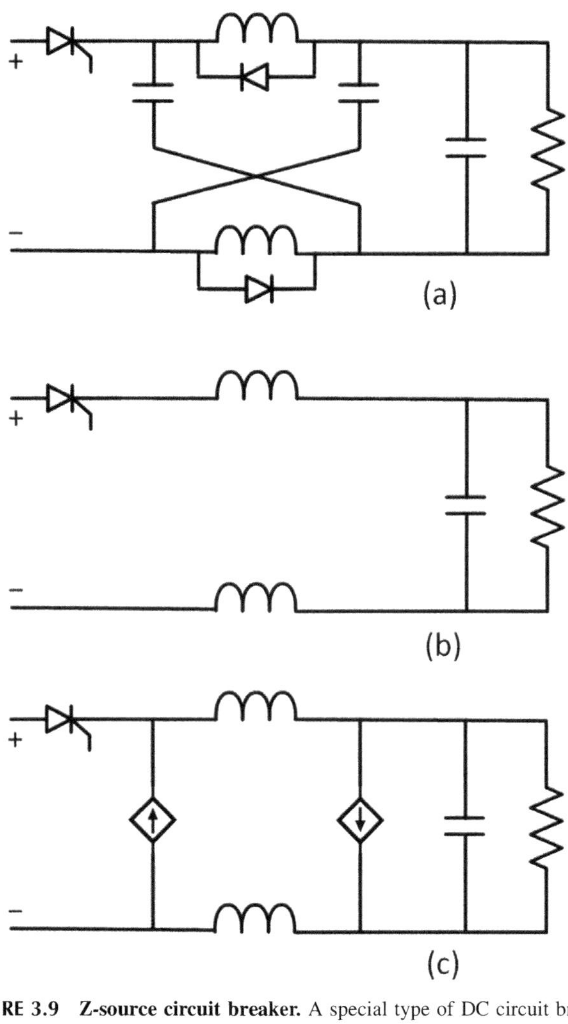


**FIGURE 3.9 Z-source circuit breaker.** A special type of DC circuit breaker.

Especially for the Z-source breaker, the immediate fault detection without sensors, its fast response and the reduced fault current flowing through the breaker with respect to the full fault current are additional advantages.

On the other hand, SSCBs suffer from high power losses on the semiconductors during normal current flows. Subsequently, an active cooling system is required, which is quite a complex and expensive system. Moreover, the need for sophisticated control makes SSCBs even more complex systems. Finally, additional cost is imposed when bidirectional SSCBs are necessary, due to the need for duplicating the semiconductor topologies and the higher on-state losses in the continuous current path. As for the Z-source breakers, the need for capacitors and a thyristor and the limited time capability for the fault current are additional disadvantages.

SSCBs can be used in all voltage levels. However, they are mostly encountered in LVDC and MVDC applications, since in HVDC level, the number and configuration of the required semiconductor topologies (i.e., the number of parallel and series switching devices), the cooling requirements, and the losses make them non-feasible and non-competitive solution techno-economically [10].

## 4.3 Hybrid circuit breakers

HCBs combine an MS in the continuous current branch with passive or active circuit components in the commutation current branch. An MOV is installed in parallel to the aforementioned branches. During short-circuit faults, the current is artificially driven to zero due to the action of the commutation current branch. Two main techniques are applied for this purpose depending on if a passive or active circuit is installed in the current commutation path [11].

In the passive circuit approach, an inductance $L$ and a capacitor $C$ are connected in series in the commutation current branch (Fig. 3.10). When the electrical contacts of the MS open due to a fault, the generated arc interacts with the parallel, resonant, $LC$ branch to produce an oscillating current $i_{osc}$ that counteracts the arcing current $i_{arc}$. The frequency of the oscillating current is given by $1/\sqrt{LC}$, and its amplitude is exponentially increasing. Zero current crossing in the continuous current branch is achieved at the time the oscillating current gets an amplitude which is equal to the fault current ($I_{DC}$). When the capacitor $C$ is being charged and the TIV exceeds the MOV conduction voltage, the energy dissipation branch is involved absorbing the produced thermal energy.

Typically, air blast or $SF_6$ mechanical switches are used in the continuous current branch of passive HCBs of resonant type (LC type), interrupting fault currents up to a few kA in tens of milliseconds. These interrupting times may be assumed relatively long for DC current breaking, whereas breaking current capability in the order of a few kA is also considered low for HVDC applications [3].

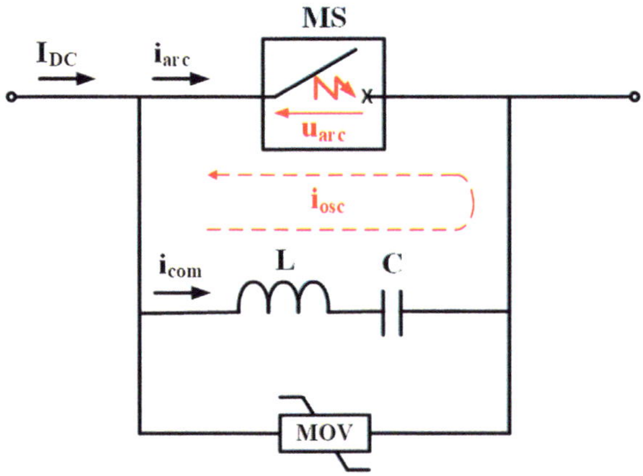


**FIGURE 3.10 Passive HCB.** Hybrid circuit breaker with passive resonance elements.

In the active circuit approach, power electronic devices are connected to the commutation current branch. Three different types of HCBs can be found in this category: (1) active resonant type, (2) natural commutation type, (3) arc-less type [5].

In active resonant HCBs (Fig. 3.11), the fault current is interrupted similarly to the passive HCBs. The difference in this case is that a pre-charged capacitor is used in the commutation current branch, while a high-speed semiconductor switch (SS) controls the capacitor discharge current [11]. The significant advantages of active resonant HCBs compared with the passive ones are the faster interruption time and the higher breaking capability. Due to the better and faster current interrupting performance, vacuum MSs are used in the continuous current branch instead of gas CBs. The additional costs related with the power electronic devices and their control are some disadvantages.

When an electronically controlled voltage source (VSC) is connected in series with the *LC* resonant circuit in the commutation current branch, a slight modification of the previously described active resonant HCB type is obtained (Fig. 3.12) [12]. The voltage source excites the oscillating current, which now takes higher amplitudes, increasing the maximum breaking capability of the circuit breaker. The frequency of the voltage source must be tuned to the frequency of the resonant *LC* circuit. With proper selection of the frequency and voltage amplitude, faster interruption is achieved.

Fig. 3.13 shows an alternative configuration of active resonant HCBs, where the pre-charged capacitor $C_1$ is connected on the LV side of an HV/LV pulse transformer [13]. In the HV side (i.e., in the commutation current branch), a second uncharged capacitor $C_2$ is connected. The discharge of the capacitor $C_1$ is controlled through a thyristor valve (SCR) with anti-parallel diode. During a fault, the thyristors are being turned on, creating an oscillating current which when superimposed to the fault current provides multiple

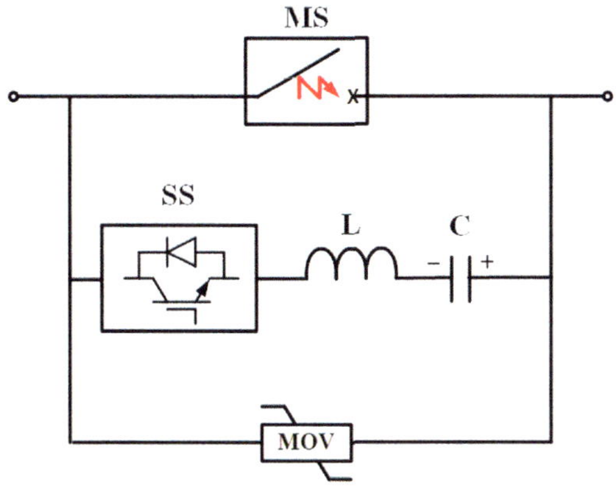


**FIGURE 3.11 Active resonant HCB.** Hybrid circuit breaker with active resonance elements.

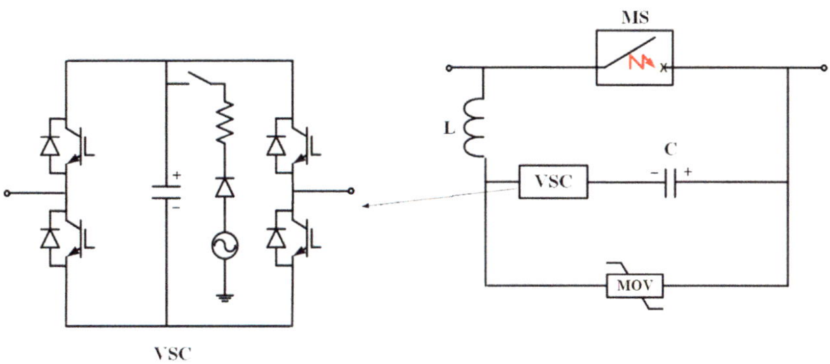


**FIGURE 3.12 Active resonant HCB with VSC.** A VSC assisted resonant current circuit breaker.

current zeros. Proper selection of the equipment parameters on both LV and HV sides and a good coupling coefficient of the pulsed transformer are critical in this configuration.

In natural commutation HCBs (Fig. 3.14), only semiconductor switches are used in the commutation current branch [14]. During a fault, these switches conduct current because of the arc voltage generated across the mechanical switch contacts, which drives the current through the commutation current branch. When the arc is totally extinguished and the dielectric strength is fully established between the contacts, the semiconductors are turned off. Then, the MOV starts conducting and clamps the voltage.

Natural commutation HCBs are mostly encountered in LVDC and MVDC applications. The main drawback of natural commutation HCBs is that the fault current is forced to flow for a significant time (i.e., for a few

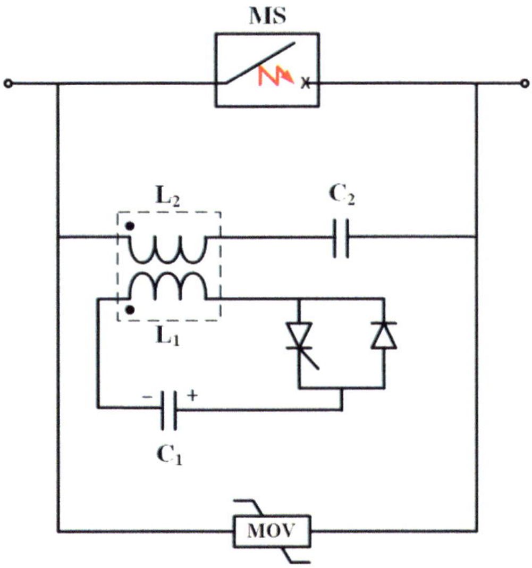


**FIGURE 3.13 Active HCB with pulse transformer.** Hybrid circuit breaker with coupled transformer and active elements in the commutation current branch.

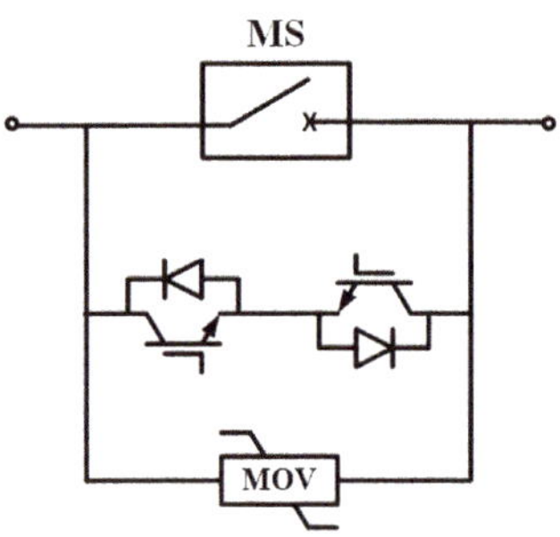


**FIGURE 3.14 Natural commutation HCB.** A DC circuit breaker with a semiconductor switch in parallel to the mechanical switch.

milliseconds) through the power semiconductors. Hence, semiconductors of larger size and cost are required compared to other types of HCBs.

Arc-less HCBs are classified in (1) HCBs based on switch-mode current injection, (2) HCBs with a load commutation switch (LCS). Switch-mode current-injection-based HCBs (Fig. 3.15) produce a current in the commutation current branch during faults, which is accurately controlled by the power electronic devices [15]. This current dynamically matches the fault current, creating the conditions for local current zero in the MS. As a result, very fast and arc-less current interruption is achieved. This type of HCBs is encountered both in MVDC and HVDC applications because of the aforementioned significant advantages.

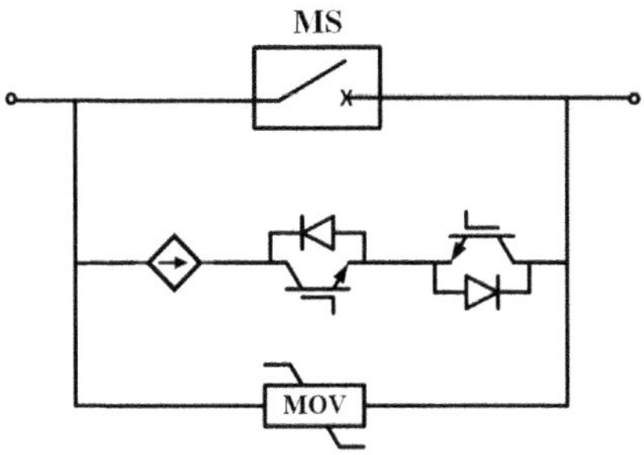


**FIGURE 3.15 Switch-mode current-injection-based HCB.** A hybrid circuit breaker with controlled counter-injection current source in the commutation current branch.

In the HCB topology of Fig. 3.16, a power electronics circuit called LCS is connected in series with the mechanical switch [16]. During faults, the semiconductors of the LCS are turned off, commutating the fault current to the commutation current branch and permitting the MS to open under zero current conditions. Once the contacts of the mechanical switch are open, the semiconductors in the commutation current branch turn-off, forcing the MOV to arrest the TIV and clamp the voltage.

There are also other HCB configurations that combine characteristics from different HCB types. In the topology of Fig. 3.17 [17], two MSs are connected in series in the continuous current branch: an ultrafast switch (UFS), which is designed to handle the full TIV, and a conventional mechanical switch (CMS), which is designed to withstand the arc voltage.

The CMS is also connected in parallel to a full-bridge commutation circuit with a pre-charged capacitor. Under non-fault (normal) conditions, the semiconductors of this bridge remain in the off-state. Hence, load current flows through the MSs, minimizing the on-state power losses during normal operation. Upon a short-circuit fault, the CMS opens, while the semiconductors of the full-bridge are turned on to conduct the current. The pre-charged capacitor discharges through the commutation inductor to create current zero in the CMS.

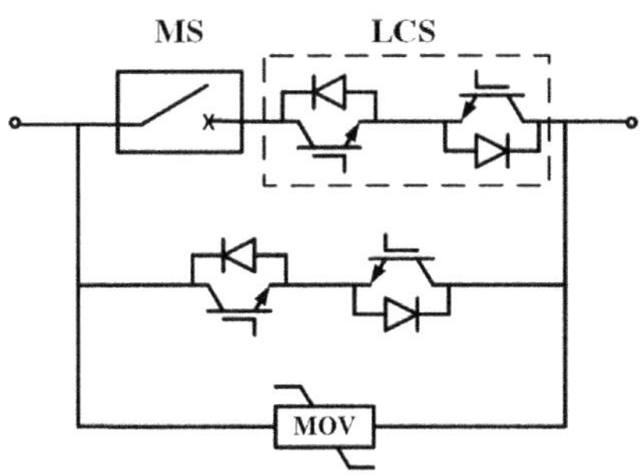


**FIGURE 3.16 HCB with an LCS.** Typical hybrid circuit breaker with a load commutation switch.

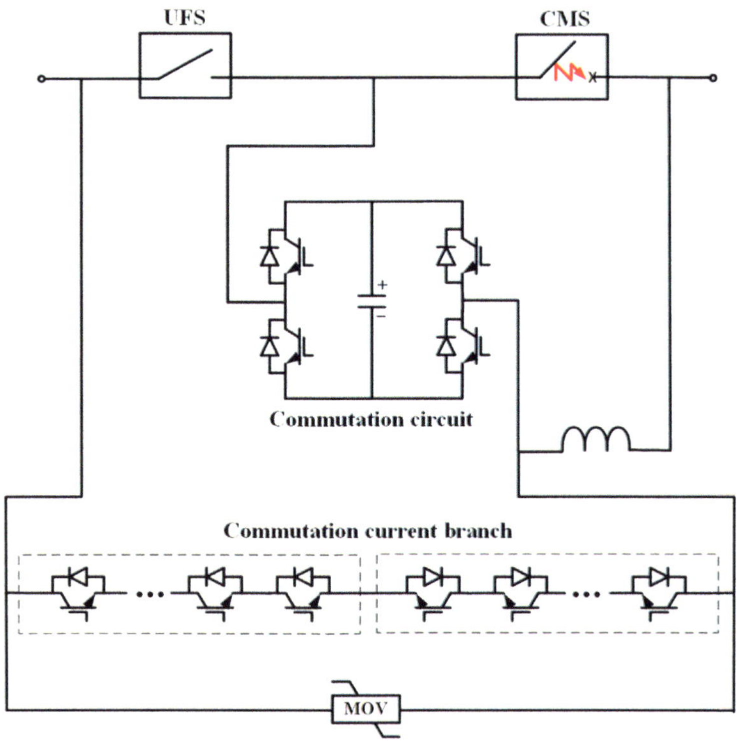


**FIGURE 3.17 HCB with parallel commutation circuits.** A special type of a hybrid circuit breaker with two mechanical switches.

After a while, the UFS opens under arcing conditions. The semiconductors in the commutation current branch are turned on, whereas the semiconductors of the bridge are turned off. The commutation circuit stops conducting, and the capacitor in the bridge starts charging. As a result, the current is commutated to the commutation current branch. When the UFS recovers its dielectric strength, the semiconductors on the commutation current branch are turned off, leaving the MOV to deal with the generated TIV.

An alternative type of HCBs is based on a gas discharge tube, which is installed in parallel with the capacitor in the commutation current branch (Fig. 3.18) [18]. During the fault and as long as the contacts of the mechanical switch are opening, the tube applies a magnetic field which is perpendicular to the electric field. The generated arc voltage commutates current to the tube which conducts until the mechanical switch fully interrupts the arc current. Then, the tube is turned off, so that the system current is commutated through the capacitor which charges until the voltage conditions drive the current through the MOV.

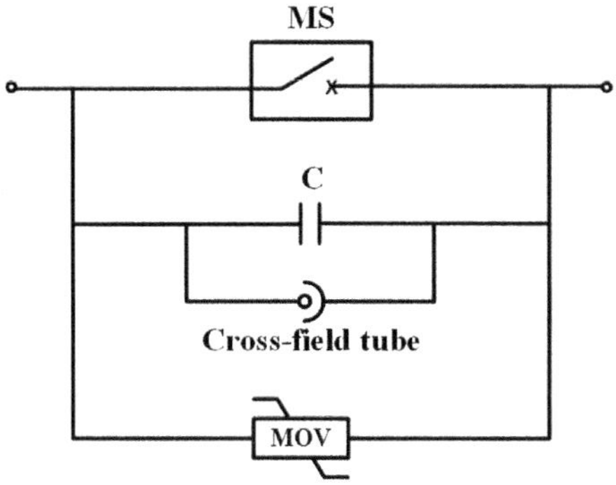


**FIGURE 3.18** **Tube switch.** Magnetic-field-based fault current elimination.

In general, active HCBs act faster than the passive ones. In addition, they show a higher breaking capability. On the other hand, more expensive and complex equipment is required in active HCBs, whereas semiconductors are exposed to high currents for a relatively long time.

### 4.4 Inherent current limiters

Some power converters have the capability to operate for a limited time under specific overcurrent conditions. They have also the capability to limit overcurrents to an acceptable level. If all power converters in a power system have such capabilities, then the expected fault currents are managed in a controlled manner.

Inherent fault current limitation eliminates the need for individual DC circuit breakers in a DC power system. Obviously, the major requirement in this case is the need for all power converters to be of this fault-tolerant type. Moreover, a power system with purely DC components is more suited for this scheme. Otherwise, intermediate power converters are required to drive all AC sources and loads, which makes system operation even more complex.

There are also power converter topologies that can totally block fault currents. Typical fault-blocking converters are the full-bridge modular multilevel converters (FB-MMCs) which provide protection for both fault current directions. Such topologies require additional semiconductors than the minimum possible and therefore they have a high cost.

## 5. Semiconductor characteristics

A wide variety of power semiconductors can be used in SSCBs and HCBs. IGBTs are a good solution because they offer a high current rating (up to 1 kA for a single unit), a high voltage rating (600–6500 V), and a high turn-off capability. The on-state resistance of a silicon IGBT is about 15–20 mΩ.

Because they are transistors, they can inherently control the current flow as well. IGCTs offer even higher voltage (between 2500 and 6500 V) and current ratings (up to 2 kA). Their on-state resistance is about 20%–50% less than the on-state resistance of IGBTs. GTOs have similar characteristics with IGCTs but require an individual, external, turn-off circuit. SCRs can operate up to 12 kV and 6 kA and offer the lowest on-state resistance. MOSFETs have an on-state resistance which increases significantly with voltage, practically not permitting their use for voltages larger than 300 V. WBG devices based on SiC or GaN (for instance, JFET or SIT) offer a significantly low on-state resistance (10 mΩ for GaN), while also they can withstand higher temperatures. On the other hand, they are expensive, and they are mostly competitive for LV applications.

## 6. Fault detection

Protection against faults in DC power systems is primarily obtained with the aid of dedicated protection devices (e.g., protection numerical relays) [10,19]. Short-circuit conditions in a DC power system are detected by these devices by sensing characteristic electrical quantities with the aid of appropriate sensors. Typical sensed quantities are voltage and current or their combination in various forms (magnitude, derivative, travelling wave). Depending on the design of the protection scheme, additional and more complex quantities (e.g., current difference, waveform shape, post-processed signals, etc.) might be utilized.

In this section, we are interested in fault detection functions that are integrated into or inclusively related with DC circuit breakers and not in the aforementioned devices (protection relay functions). Integrated fault detection functions provide inherent protection to DC circuit breakers against short-circuit faults. In the following, the basic methods applied by DC circuit breakers to detect short-circuit fault are presented.

The simplest way to sense DC current is by using a shunt resistor or Hall-effect sensors [20]. In the first method, a low ohmic resistor is connected in the continuous current branch. The current is indirectly measured by measuring the voltage drop on the resistor. Although the resistance is small, power losses are produced, whereas also the transient response of the resistor during faults interacts with the remaining DC loop transients. In addition, there is no galvanic isolation of the resistor with the protected circuit, and therefore, the latter must be isolated from the protection unit.

Hall-effect sensors base their operation on the fact that when a current flows in a circuit across a vertical magnetic field, a voltage is induced that is perpendicular to the magnetic field and the current [20]. Commonly, zero flux DC current transformers are used for measuring DC current in practical applications. The primary circuit DC current is flowing through the primary winding of the transformer, which is wound around a magnetic core. The generated magnetic field in the core induces a voltage at the Hall-effect sensor output, which is proportional to the measured input current.

An alternative way to sense a fault in a DC power system is to apply the de-saturation technique [6]. This technique is only suited to SSCBs or HCBs, which comprise power electronic devices. Short-circuit conditions in the DC circuit are detected when the voltage across the semiconductor switches (i.e., the voltage between the drain and the source terminals) diverts from the low on-state value to the DC bus voltage following the power output curve. Fast and reliable fault detection is achieved in this way, but the exact overcurrent cannot be measured. In addition, the protection circuit is not isolated from the protected DC circuit.

In commercial power electronic devices, gate drivers support the so-called de-saturation (DESAT) protection function. Again, the protection principle applied is the de-saturation of the semiconductor switch, but in this case the DC current is directly measured. If the latter exceeds a predefined threshold, a short-circuit fault is detected.

## References

[1] Low-voltage Switchgear and Controlgear—Part 2: Circuit-Breakers Standards Australia, 2005, 2005.

[2] IEEE Standard for Electrical Power System Device Function Numbers, Acronyms, and Contact Designations, IEEE Standard, 2022, 2022.

[3] N.A. Belda, HVDC Circuit Breakers—Test Requirements . Methods and Circuits, 2021, 2021.

[4] N.A. Belda, R.P.P. Smeets, Test circuits for HVDC circuit breakers, IEEE Trans. Power Deliv. 32 (1) (2017) 285–293, https://doi.org/10.1109/TPWRD.2016.2567783.

[5] I.C. Kizilyalli, Z.J. Shen, D.W. Cunningham, Direct Current Fault Protection: Basic Concepts and Technology Advances Power Systems, Springer International Publishing, 2023, https://doi.org/10.1007/978-3-031-26572-3.

[6] W.Y. Kong, Review of DC Circuit Breakers for Submarine Applications, Australian Government, Maritime Platforms Division, DSTO Defence Science and Technology Organisation, 2012, 2012.

[7] Abb Circuit Breakers For Direct Current Applications Abb Document: 1SXU210206G0201.

[8] X. Pei, O. Cwikowski, D.S. Vilchis-Rodriguez, M. Barnes, A.C. Smith, R. Shuttleworth, A Review of Technologies for MVDC Circuit Breakers IECON 2016—42nd Annual Conference of the IEEE Industrial Electronics Society.

[9] K.A. Corzine, R.W. Ashton, A new Z-source DC circuit breaker, IEEE Trans. Power Electron. 27 (6) (2012) 2796–2804, https://doi.org/10.1109/TPEL.2011.2178125.

[10] S. Zheng, R. Kheirollahi, J. Pan, L. Xue, J. Wang, F. Lu, DC circuit breakers: a technology development status survey, IEEE Trans. Smart Grid 13 (5) (2022) 3915–3928, https://doi.org/10.1109/tsg.2021.3123538.

[11] A. Shukla, G.D. Demetriades, A survey on hybrid circuit-breaker topologies, IEEE Trans. Power Deliv. 30 (2) (2015) 627–641, https://doi.org/10.1109/TPWRD.2014.2331696.

[12] L. Ängquist. S. Nee. T. Modeer. A. Baudoin. S. Norrga. N.A. Belda, IET conference publications 10.1049/cp.2019.0034 9781839530074 751 institution of engineering and

technology Sweden design and test of VSC assisted resonant current (VARC) DC circuit breaker (2019). http://www.ietdl.org/CP 2019.

[13] L. Liu, Z. Yuan, H. Xu, L. Chen, J. He, P. Yuan, Y. Cen, Design and test of a new kind of coupling mechanical HVDC circuit breaker, IET Gener., Transm. Distrib. 13 (9) (2019) 1555–1562, https://doi.org/10.1049/iet-gtd.2018.5658.

[14] J. Magnusson, A. Bissal, G. Engdahl, J.A. Martinez-Velasco, IEEE PES innovative smart grid technologies conference Europe 10.1109/ISGTEurope.2014.7028737 9781479977208 January IEEE computer society Sweden design aspects of a medium voltage hybrid DC breaker, (2015). https://ieeexplore.ieee.org/xpl/conhome/1800214/all-proceedings2015.

[15] Y. Zhou, Y. Feng, N. Shatalov, R. Na, Z.J. Shen, An ultraefficient DC hybrid circuit breaker architecture based on transient commutation current injection, IEEE J. Emerg. Select.Topics Power Electr. 9 (3) (2021) 2500–2509, https://doi.org/10.1109/JESTPE.2020.2983354.

[16] M. Callavik, A. Blomberg, J. Häfner, B. Jacobson, The Hybrid HVDC Breaker ABB Grid Systems Technical Paper, 2012, p. 2012.

[17] K. Kanaya, Y. Ohda, Y.K.K. Hashimoto, T. Hoshino, T. Ishiguro, Koyama. Y, A.D.M. K Takimoto, CIGRE-IEC 2019 Conference on EHV and UHV (AC and DC) UFD with High Arc Resistance and Ielectric Recovery Characteristics for Current Commutated Hybrid DCCB.

[18] A.L. Courts, J.J. Vithayathil, N.G. Hingorani, J.W. Porter, J.G. Gorman, C.W. Kimblin, A new DC breaker used as metallic return transfer breaker, IEEE Trans. Power Apparatus Syst. PAS-101 (10) (1982) 4112–4121.

[19] L. Hallemans, G. Van Den Broeck, S. Ravyts, M.M. Alam, M.D. Vecchia, P. Van Tichelen, J. Driesen, Fault identification and interruption methods in low voltage DC grids-a review, in: IEEE 3rd International Conference on DC Microgrids, ICDCM 2019 10.1109/ICDCM45535.2019.9232856 9781728134918, Institute of Electrical and Electronics Engineers Inc., Belgium, 2019. http://ieeexplore.ieee.org/xpl/mostRecentIssue.jsp?punumber=9171388.

[20] D. Tzelepis, V. Psaras, E. Tsotsopoulou, A. Mirsaeidi, Q. Dysko, X. Hong, S.M. Dong, V. Blair, V. Nikolaidis, G. Papaspiliotopoulos, G. Fusiek, P. Burt, C. Niewczas, Booth, voltage and current measuring technologies for high.

Chapter 4

# Switchgear faults and remedies

**Krishna Murari**[1] **and Vishal Kumar Gaur**[2]
[1]*The University of Toledo, Department of Electrical Engineering & Computer Science, Toledo, OH, United States;* [2]*Motilal Nehru National Institute of Technology Allahabad, Electrical Engineering Department, Allahabad, Uttar Pradesh, India*

## 1. Introduction to switchgear and its faults

The reliable and efficient operation of a power system is essential for the smooth functioning of modern society. For ensuring the safe and continuous supply of electrical energy, power system switchgear plays a crucial role. Switchgear is a general term referred to various equipment that are basically used for switching and protection functions. These components includes busbars, disconnector switches, circuit breakers, relays, lightning arresters, instrument transformers, etc. These are essential components of any power system. Hence, the probability of faults in switchgear itself cannot be ignored and is of great concern for reliable operation of whole system. Switchgear faults include a range of anomalies, malfunctions or failures within the intricate network of components that collectively constitute switchgear. These faults, if left unaddressed, can result in a cascade of consequences, including disruptions in the electrical supply, damage to critical equipment, and in certain instances, the emergence of safety hazards.

## 2. Types of switchgear faults

Understanding the types of switchgear is essential as each type has its characteristics, affecting fault probability and maintenance strategies. Common switchgear types include air-insulated switchgear (AIS), gas-insulated switchgear (GIS) and hybrid switchgear. The nature or type of fault and its probability of occurrence may depend upon the specific type of switchgear deployed [1]. This comprehension serves as the bedrock for implementing effective maintenance strategies, conducting thorough troubleshooting and pre-emptively addressing issues before they escalate. In this context, an exploration of some common switchgear faults becomes instrumental.

**Switchgear Design, Operation, and Maintenance Using Industry Standards**
**https://doi.org/10.1016/B978-0-443-24718-7.00004-4**

Switchgear faults are classified in three major categories; mechanical faults, electrical faults and environmental issues.

## 2.1 Mechanical faults

Most of the electrical switchgear utilized in field require some kind of mechanical operating mechanism for particular task or operation. For example, high-voltage circuit breakers often use a spring-operated mechanism for fast and reliable operation. When a fault occurs, the stored energy in the springs is rapidly released to open the contacts, interrupting the current flow. Disconnect switches use a rotary mechanism to physically disconnect or connect electrical circuits. They often feature a rotating blade or arm that makes or breaks contact with fixed terminals. Manual or motor-operated mechanisms can be used to actuate the switch, allowing for remote operation when necessary. Disconnect switches are commonly used in substations and transmission lines for isolating sections of the electrical grid during maintenance or in case of emergencies. Another example can be grounding switches that are also known as earthing switches. They are used to ground or connect electrical equipment to the earth for safety purposes. High-voltage grounding switches often employ pantograph mechanisms, which consist of articulated arms that extend or retract to make or break the grounding connection. These mechanisms allow for smooth and reliable operation, even under high-voltage conditions. Busbar couplers is also an example in this series. It is utilized to connect or disconnect sections of electrical busbars in substations or switchyards. Mechanical interlocking mechanisms are commonly employed to ensure proper alignment and engagement of the busbar couplers before making electrical connections.

Faults associated particularly with these mechanical subsystems of electrical switchgear are discussed under the category of mechanical faults. Misalignment and mechanical wear represent significant challenges in the realm of mechanical faults, contributing to potential disruptions in the operational efficiency and longevity of machinery. This section delves into the intricacies of these mechanical issues, shedding light on their manifestations and possible consequences. Misalignment occurs when components are not properly positioned or oriented according to their intended design. This deviation from the specified alignment parameters can lead to a cascade of adverse effects, impacting the overall functionality and reliability of the mechanical system. In generalized way, we can say that the misalignment or mechanical binding of moving parts within the switchgear mechanism can inhibit proper operation and lead to jamming or sticking. Misaligned components may induce higher levels of friction between moving parts, resulting in increased heat generation. This can accelerate wear and compromise the lubrication system. Considering the case of mechanical wear, it is a natural consequence of friction and abrasion between moving components within a switchgear. Understanding the facets of mechanical wear is crucial for implementing effective maintenance strategies and extending the operational lifespan of machinery. Further, we can classify the mechanical wear as abrasive wear, adhesive wear and fatigue wear. Abrasive wear is mainly caused by the presence of abrasive particles in the lubricating system or contact surfaces, resulting in surface erosion. On the other hand, adhesive wear occurs when two surfaces in relative motion adhere to each other, leading to the transfer of material and the formation of wear debris, whereas the fatigue wear is gradual wear and tear due to repeated cyclic loading and unloading of components, resulting in cracks and material fatigue. There are many switchgears, which we can easily recognize in our power system, performing some type of mechanical operation, i.e., circuit breaker with operating mechanism, disconnector switch, earthing switch, motorized mechanism. Taking the example of a circuit breaker, adhesive wear occurs between fixed and moving contacts. Further, many switchgear mechanisms rely on springs to provide the necessary force for operation. A failed spring can lead to improper operation of the switchgear, such as failure to close contacts or insufficient force to trip a circuit breaker. Therefore, it is also important to note that the springs and other mechanical parts of operating mechanisms associated with switchgear operating mechanism can experience fatigue wear due to repeated operation. These wear may affect and lead to degraded performance, increased contact resistance and eventual failure. Similarly, disconnector has repeated operation between fixed contact (female contact) and moving contact (male contact), as shown in Fig. 4.1, which causes adhesive wear between them. These wear

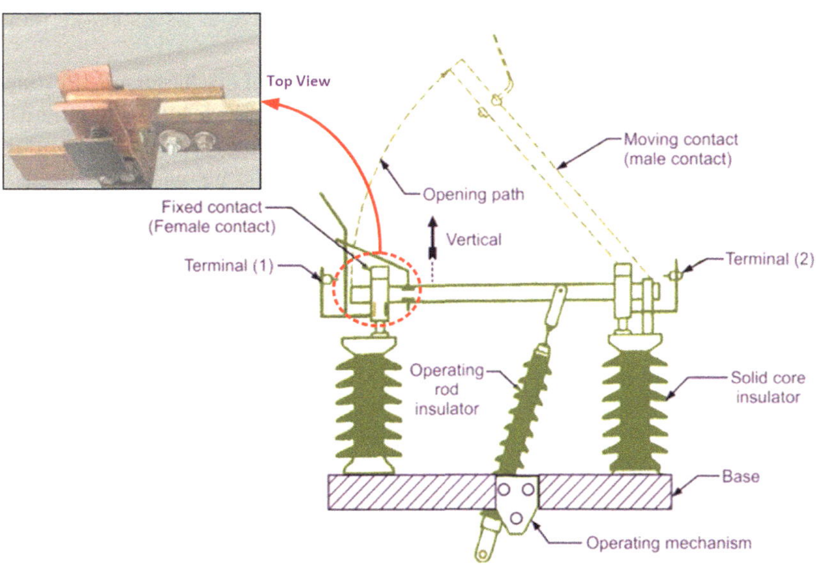


**FIGURE 4.1** Pictorial view of fixed and moving contact of disconnector switch.

effects can lead to increased friction, reduced contact pressure, increased contact resistance and eventual failure of the switchgear components.

## 2.2 Electrical faults

Electrical faults in switchgear pose significant challenges to the reliable operation of power systems. Understanding these faults and their consequences is crucial for effective maintenance and fault management. Electrical faults in switchgear, such as arc faults and insulation failures, pose significant challenges to power system reliability. Each type of switchgear (AIS, GIS, hybrid) has unique features influencing how these electrical faults manifest and should be addressed.

### *2.2.1 Insulation failures*

Insulation failures compromise the ability of switchgear to withstand voltage stresses and maintain proper electrical isolation. Insulation can degrade due to heat, chemicals or other factors including environmental conditions. Proper insulation is crucial for preventing the impact of short-circuit or overcurrent and ensuring the safe operation of the power system. Further, the impact of environmental condition is also discussed in other section of this chapter. Insulation failures result in reduced dielectric strength, making the switchgear susceptible to breakdowns and flashovers. This can lead to equipment malfunction and pose a significant risk to the integrity of the power system. Insulation failures may necessitate the replacement of damaged components or, in severe cases, entire switchgear units. Regular insulation testing and maintenance are essential to identify and address potential issues before they escalate [2].

For metal-clad switchgear or AIS, surface electrical discharge, manifesting as corona and tracking, has been a significant factor leading to failures. Corona, characterized by a faint light surrounding an electrical conductor due to air ionization, poses a threat when nitrogen in the air undergoes breakdown. The resulting nitric acid, formed in the presence of moisture, has deleterious effects on insulating materials and metallic components, potentially causing heat-related issues. Notably, corona formation is insidious as it occurs quietly, without causing a flashover, when nearby air is strained beyond its ionization threshold by electrical stress. The most perilous aspect of corona discharge is its occurrence without a visible flashover. The process involves the charging of air layers within insulation when electrical stress exceeds its insulative properties. Spatial factors play a crucial role in corona occurrence, involving small air spaces between conductors, insulation boards, switchgear cabinet components, etc. This may result from conductors being tie-wrapped together, touching insulators, conduit, or cabinet edges, non-shielded cables in contact with grounded surfaces and busbars in close proximity to insulating materials [3].

Further, talking about current transformers (CT), shown in Fig. 4.2, the insulating material may harbour high humidity, leading to creeping discharge on its bushing surface. This situation is more prevalent when terminal leads are discharged along the surface of the pad. Although CT is not purely switchgear, but they are oftenly associated with switchgear and specifically used with switchgear for measurement and protection.

### 2.2.2 Arc faults

Arc faults in switchgear are one of the most significant hazards, mostly due to the amount of energy the arc may contain and potential consequences [4]. Arcs occur when electric current strays from the designed route, generating heat, light and pressure. The risks and management of arc faults are different for types of switchgear, such as AIS and GIS. AIS utilize air as the primary dielectric medium. The components are physically spaced apart to prevent fault conditions, and the switchgear is typically housed in large metal enclosures. Being exposed to the atmosphere, AIS is prone to environmental effects such as humidity, pollution and temperature fluctuations, which can degrade insulation and component integrity. Arc faults can result from exposure to conductive parts due to any physical injury during maintenance, operations or external occurrences. Mechanical elements of switchgears such as circuit breakers and disconnect switches wear out over time as a result of frequent on-off cycles, which impose electrical stresses that may damage insulation and contacts. Additionally, insulators dielectric strength degrades with ageing leading to less efficaciousness and hence, more arcing occurrences.

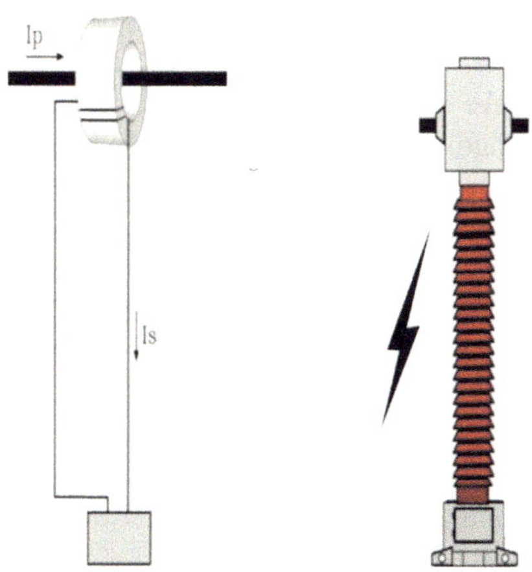


**FIGURE 4.2** Creeping discharge through bushing surface in a current transformer.

For GIS, the integrity of SF6 gas needs to be maintained at all times. Any reduction in the purity or density, whether caused by leak or contamination, drops its insulating properties, potentially leading to arc faults. Although SF6 is a highly effective as an insulator, its effectiveness reduces if it leaks. Further, defects at sealing points, bushings, or the containment of gas itself can initiate arc faults. The high-speed switching operations, typical in GIS, can also trigger arc faults in some cases, i.e., due to transient conditions. Arc faults in GIS can inflict severe damage on the internal workings of the switchgear, necessitating costly repairs. Moreover, arcing can breakdown SF6 into toxic byproducts, which pose health risks during maintenance efforts [5].

## 2.3 Environmental faults

Environmental faults in switchgear refer to the impact of external conditions on the performance and integrity of the equipment. Switchgear, being an integral part of power systems, is often exposed to various environmental factors that can lead to degradation and faults. Understanding these environmental faults is crucial for implementing preventive measures and ensuring the longevity of switchgear.

### *2.3.1 Corrosion and degradation*

Corrosion is a common environmental fault that affects the metal components of switchgear, such as enclosures, busbars and support structures. It occurs due to the reaction of metal surfaces with environmental elements, such as moisture, oxygen and pollutants.

Corrosion can weaken the structural integrity of switchgear components, leading to reduced mechanical strength. It may also affect the electrical conductivity of busbars, leading to increased resistive losses and potential overheating.

For example, current transformers (CTs) are specifically designed for outdoor use; however, there have been instances where certain materials have proven inadequate for this purpose or have aged prematurely under normal operating conditions [6]. Oil-filled freestanding CTs typically feature a porcelain body, providing protection against environmental factors such as rain, dust and heat. Many manufacturers utilize stainless steel bolts and top covers in their assembly processes. Additionally, the expansion bellows of modern transformers are also constructed from stainless steel. Despite its durability, stainless steel is prone to rusting in marine environments due to high salinity and humidity levels. Over time, this corrosion can lead to hot spots and the occurrence of corona discharge activity within the components.

The nature of failures is explosive and usually engulfs the control cabinet and much of the wiring associated with these controls. As shown in Figs. 4.3 and 4.4,

FIGURE 4.3 Effects of environmental conditions in high-voltage CTs [6].

the explosive nature of the CT failures propels porcelain products long distances and results in extensive damage to nearby equipment.

### 2.3.2 Effects of temperature

Switchgear is exposed to a wide range of temperatures, from extreme cold to high heat, depending on the geographical location and climate. Temperature extremes may impact the performance of insulation materials and affect the mechanical properties of switchgear components. Cold temperatures can lead to increased viscosity of lubricants, affecting the operation of moving parts.

FIGURE 4.4 Violent failure of a 400 kV current transformer (damaged switchgear) [6].

High temperatures can accelerate the ageing of insulation materials and may lead to insulation breakdown.

Weather conditions extremities may affect the high- and medium-voltage substations. For example, in northern countries, some of the most prevalent severe climate conditions that impact AIS occur during winter, including low temperatures, heavy snowfall and ice storms. These conditions may impacts on the components of AIS switchgears, such as circuit breakers and disconnector switches, often resulting in equipment failures, alarms and sporadic blackouts. It is essential to individually outline these conditions and explore available mitigation strategies for each of them.

The majority of electrical equipment intended for outdoor use adheres to international standards are designed to operate within specified parameters, i.e., between temperatures of +30 and +40°C. To ensure the reliable performance in the given range of operating conditions, manufacturers have to conduct type tests accordingly for such switchgear. However, many utilities operate in northern countries where temperatures can be beyond the range for it is typed test, i.e., below negative 30°C, and occasionally even lower than negative 50°C, during certain period of the year. Switchgear exposed to such extremely low temperatures may experience various effects, influenced by factors such as its construction, insulating medium, lubricants utilized for moving parts, gaskets, seals and fabrication materials. Regarding GIS switchgear, gas-filled equipment may be vulnerable to leaks if the weight of snow and ice distorts the equipment, compromising the seals. This risk can particularly affect GIS outdoor bushings or other gas-filled equipment, such as surge arresters. One potential mitigation strategy could involve installing snow protection covers on the insulators to prevent the accumulation of heavy snow and ice. Similarly, at low temperatures, different gaskets and seals may be impacted. Depending on their manufacturing materials, they may lose elasticity or shrink, potentially leading to leaks of insulating fluids (such as oil, compressed air, SF6 or other gases). These leaks can result in environmentally harmful emissions, trigger low-pressure alarms or even disrupt operations if the insulating fluid levels fall below acceptable thresholds. Additionally, very low temperatures can cause moisture ingress and corrosion of internal components, further affecting the gaskets of control cabinets.

For instance, circuit breakers relying on pneumatic mechanisms may encounter issues if the compressed air lacks sufficient dryness. In extremely low temperatures, moisture can condense and freeze, potentially causing blockages in air pipes and resulting in operational failures. Furthermore, ground freezing can adversely impact the foundations supporting AIS equipment. The extreme cold can expedite erosion of the footings, leading to movement that may affect the operation of components like HV disconnect switches. Moreover, very low temperatures can affect insulating fluids. For example, increase in viscosity of insulating oil reduces the circulation within reactive equipment. It causes the challenges for moving parts in switching equipment. Hence, this slows down the interrupting process. Additionally, the outdoor SF6 circuit breakers are typically operate at pressures around 480 kPa (70 psi) that begins liquifying around −32°C and causes triggering of low-gas alarms. Hence, it influences the interrupting performance of the breakers.

SF6-insulated equipment is sensitive to substantial temperature fluctuations, which can trigger alarms or even result in tripping due to changes in gas pressure. Similarly, transformers and potential devices are susceptible to these temperature variations, leading to significant drops in oil levels during prolonged cold spells. It is crucial to regularly check and maintain equipment oil levels at the manufacturer's recommended level to mitigate potential issues. Additionally, site access is major challenge during the snowdrifts and can impact the interior of the substation if blown against primary equipment. The increase in height of the perimeter fence may be a possible countermeasure to address such issue.

## 3. Remedial actions for various types of faults

### 3.1 Remedial actions for mechanical faults

Mechanical faults in switchgear can be addressed through various remedial actions aimed at restoring proper functionality and reliability:

#### *3.1.1 Misalignment remediation*

To address misalignment issues in switchgear, the following remedial actions can be taken:

- Implementation of preventive maintenance schedules to monitor and address misalignment issues before they escalate.
- Regular alignment checks using precision tools can help to detect misalignment issues on early stage.
- Implementing vibration analysis techniques allows for the identification of abnormal vibration patterns associated with misalignment.
- In some of the critical switchgear application, monitoring temperature differentials using thermal imaging can also reveal areas prone to misalignment-induced heat generation.
- Lubrication of moving parts to reduce friction and prevent binding or sticking.
- Replacement of worn or damaged components contributing to misalignment, such as bearings, shafts or couplings.

#### *3.1.2 Mechanical wear remediation*

Remedying mechanical wear in switchgear involves the following corrective actions:

- Replacement of worn out or damaged elements, such as contacts, springs, or operating mechanisms, to restore proper functionality.
- Lubricating and treating the surface of mechanical parts so as to minimize friction as well as rates at which they wear off.
- Design improvements with an aim of counteracting problems related to wear within switchgears.
- Taking up inspection methods based on predictions which may involve monitoring conditions and analyzing particles produced by wearing off, to detect and address mechanical wear proactively.

## 3.2 Remedial actions for electrical faults

Electrical faults in switchgear can have serious implications for the reliability and safety of power systems. Implementing effective remedial actions is crucial to address these faults and prevent their recurrence. Here are some key remedial actions for addressing electrical faults in switchgear:

### *3.2.1 Insulation failure remediation*

- Conduct regular inspections to check for any signs of insulation degradation caused by heat, environmental factors or chemical exposure.
- Implement routine maintenance plan to replace or repair insulation materials with signs of wear or damage.
- Measures to control outside temperature, humidity and pollution levels, where switchgear is located, to minimize the risk of insulation failure.
- Choose durable high-quality insulating materials to withstand environmental stress and chemical degradation.
- Consider retrofitting existing switchgear with advanced insulation technologies that offer better performance and longer lifespans.
- Ensure implementation of spacing and insulation guidelines to avoid corona discharge, particularly in high-voltage system.
- Protect against the formation of nitric acid and subsequent material degradation by utilizing the corona-resistant materials and coatings.
- For components such as current transformers, ensure that insulating materials are kept dry and well maintained to avoid creeping discharges.

### *3.2.2 Arc faults remediation*

- Install advanced arc fault detection systems that can quickly identify and isolate arc faults before they can cause significant damage.
- Implement monitoring systems to detect SF6 gas leak and ensure gas integrity is maintained to reduce the chances for an arc faults.
- Check on a regular basis purity as well as density of gas to achieve desired insulation properties against arc quenching actions.

- For AIS, ensure adequate physical spacing between components and use arc-resistant designs to minimize possibility of arc initiation due to physical damage or environmental conditions.
- For GIS, enhance the design of seals, bushings, and gas containment systems to prevent defects that could lead to arc faults.
- Equip switchgear with fast-operating protection devices that can immediately isolate faulted sections to prevent the spread of damage.
- Implement protective relays and circuit breakers designed to handle high-speed switching operations and electrical stresses.
- Train maintenance personnel on the risks associated with arc faults and the proper handling of SF6 gas, including safety measures for dealing with toxic by-products.

## 3.3 Environmental faults remediation

To address environmental faults affecting switchgear, the following remedial actions can be implemented:

- Installation of protective enclosures or shelters to shield switchgear from adverse environmental conditions such as corrosion, moisture or temperature extremes. Fig. 4.5 shows the installed of protective enclosures and climate temperature control device for circuit breaker.
- Application of corrosion-resistant coatings or materials on switchgear components to enhance durability and longevity.
- Implementation of climate control systems to regulate temperature and humidity levels within switchgear enclosures.
- Periodic inspection and cleaning of switchgear to remove contaminants and prevent corrosion or degradation of components.

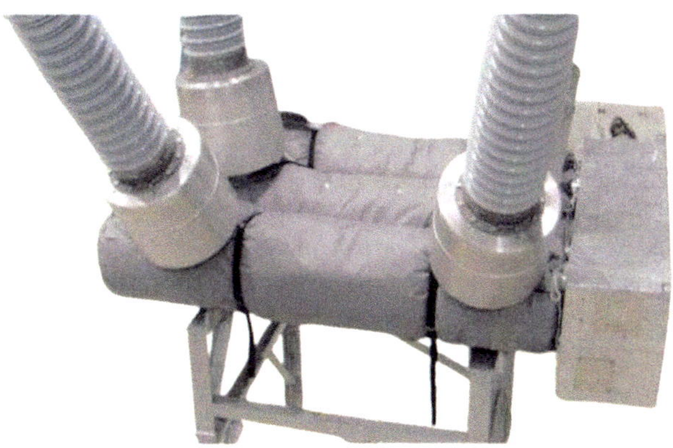

**FIGURE 4.5** Circuit breaker with complete blanket enclosure for interrupter tank.

## 4. Conclusion

This chapter has provided a comprehensive exploration of switchgear faults and their associated remedies. Switchgear faults, ranging from electrical to mechanical and environmental, can significantly impact the reliability and safety of power systems. Understanding the root causes, manifestations, and consequences of these faults is essential for implementing effective maintenance strategies and preemptively addressing issues before they escalate. The chapter has highlighted various types of switchgear faults, including corona discharge, short circuits, overcurrents, insulation failures, misalignment, mechanical wear, corrosion and degradation. Moreover, the chapter has outlined remedial actions for addressing switchgear faults. Overall, by delving into the intricacies of switchgear faults and their remedies, this chapter serves as a valuable resource for engineers and practitioners involved in the maintenance and operation of power systems.

## AI disclosure

During the preparation of this work the author(s) used ChatGPT to refine the content's grammar and appearance. After using this tool/service, the author(s) reviewed and edited the content as needed and take(s) full responsibility for the content of the publication.

## References

[1] S. Ishak, et al., Fault classification system for switchgear cbm from an ultrasound analysis technique using extreme learning machine, Energies 14 (19) (2021), https://doi.org/10.3390/en14196279.

[2] P. Gill, Electrical Power Equipment Maintenance and Testing, second ed, Taylor and Francis, United Kingdom, 2009.

[3] J. Brady, L. Thermographer, Corona and tracking conditions in metal-clad switchgear case studies, (2006). Available at: https://www.irinfo.org/articleofmonth/pdf/article_8_1_2006_Brady.pdf. (Accessed 2006).

[4] H. Weichert, et al., On partial discharge/corona considerations for low voltage switchgear and controlgear, in: Electrical Contacts, Proceedings of the Annual Holm Conference on Electrical Contacts, 2018-October, 2018, pp. 246–253, https://doi.org/10.1109/HOLM.2018.8611670.

[5] Ambo, et al., AC breakdown behavior of SF6/N2 gas mixtures under non-uniform field electrode configurations, Telkomnika (Telecommunication Computing Electronics and Control) 17 (5) (2019) 2451–2456, https://doi.org/10.12928/TELKOMNIKA.v17i5.12806.

[6] Reyes, et al., Abnormal failure rate on high voltage current transformers affected by environmental conditions, in: Conference Record of IEEE International Symposium on Electrical Insulation, 2008, pp. 247–251, https://doi.org/10.1109/ELINSL.2008.4570321.

Chapter 5

# Design considerations for relay and circuit breaker

**Sandhya Chandran and Ramakrishna Gokaraju**
*Department of Electrical and Computer Engineering, University of Saskatchewan, Saskatoon, SK, Canada*

## 1. Introduction

Power system protection strategies are implemented to ensure continuous operational reliability and safeguard equipment and personnel from various electrical events. With electrical networks becoming increasingly complex, due to renewable energy resources (RERs), the need for reliable protection devices has grown significantly. Relays and circuit breakers (CB) are the most essential protection devices. They work together to detect faults and isolate the faulted network to interrupt the power flow during any abnormal electric conditions.

Relays are intelligent devices that continuously monitor network current, voltage, frequency and phase angle. During any abnormal operating conditions, the relays detect and initiate predefined protective actions by sending a command to CB or any other interrupting devices. Relays ensure selective fault isolation that minimizes system disruptions by isolating only the faulted areas and preventing cascading failures, thereby maintaining the overall stability of the grid. Modern numerical and digital relays incorporate advanced algorithms for precise fault detection and coordination with faster response times which are more useful for RERs integrated networks.

Circuit breakers are electromechanical devices that provide manual or automatic network isolation by physically interrupting the current flow in the event of overcurrent, shortcircuit or ground faults. Modern breakers are equipped with electronic or microprocessor-based trip mechanisms, which facilitate precise fault detection and configurable protection settings. CBs can be integrated with intelligent relays, enabling dynamic coordination and enhanced selectivity in efficient fault management. With modern communication protocol-enabled breakers, distance or differential protection can be

**Switchgear Design, Operation, and Maintenance Using Industry Standards**
**https://doi.org/10.1016/B978-0-443-24718-7.00005-6**

performed by adopting fast and selective tripping schemes. Also, recently with the innovations in arc-flash detection and current limiting technologies, CBs are finding their effective use in extremely high-energy applications.

## 2. Circuit breaker fundamentals

A circuit breaker is a mechanical switch that automatically interrupts the flow of electrical current to protect a circuit from damage due to overcurrent, such as from overloading or short circuits. It operates by sensing fault conditions, then opening its contacts to break the circuit using stored mechanical energy (e.g., springs or compressed air) or by relying on electromagnetic or thermal expansion mechanisms. Once the contacts separate, an electrical arc forms, which must be extinguished to prevent damage. Various methods are used for arc extinction, including cooling the arc, increasing its length, air blasting, reducing its cross-section or splitting it into smaller arcs. Circuit breakers also use different mediums, such as air, vacuum, oil or insulating gases such as SF6, to quench the arc safely. For AC breakers, the arc is extinguished at natural zero current points, while DC breakers may use capacitors and inductors to create artificial zero currents. Based on their working principle, CBs can be classified as follows:

- *Air circuit breaker (ACB)*
  Air circuit breakers (ACBs) are electrical protection devices used primarily in low-voltage applications, typically up to 1000 V AC, to protect electrical circuits from overloads, short circuits and ground faults as below in Fig. 5.1. They operate by using air as the medium to extinguish the arc that forms when a fault is interrupted, both plain air or blast of pressured air. ACBs are available in fixed and draw-out designs, with current ratings typically ranging from 800 A to 6300 A and short-circuit breaking

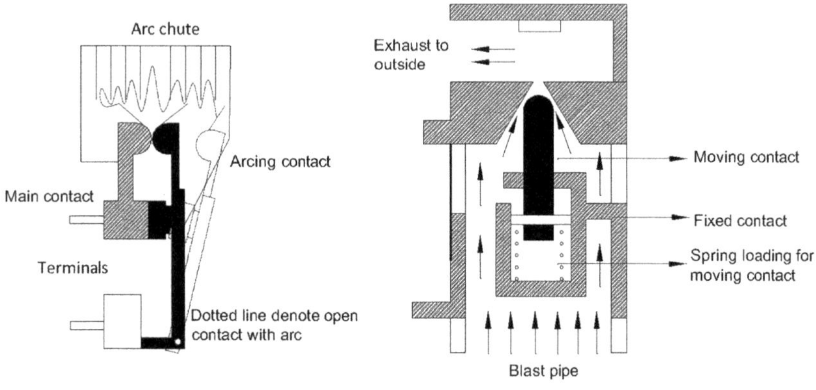


**FIGURE 5.1** ACB breaking philosophy: (A) plain ACB; (B) axial blast ACB.

capacities from 35 to 100 kA. Modern ACBs are equipped with electronic trip units that allow precise protection settings, including adjustable overload protection, short-circuit protection and ground fault detection. ACBs are preferred for their high reliability, durability and ability to handle frequent switching operations. Additionally, they often feature integrated communication and monitoring systems, making them suitable for applications in industrial plants, commercial buildings and power distribution networks. Their ability to operate in harsh conditions, combined with maintenance friendly designs, ensures effective protection and longevity in electrical installations.

- *Miniature circuit breaker (MCB) and moulded case circuit breaker (MCCB)*
  An MCB is used in low current applications, typically below 100 A, to protect circuits from overloads and short circuits in homes and commercial buildings, as shown in Fig. 5.2. It operates automatically by opening the circuit when the current exceeds its rated limit. MCBs use a thermal mechanism (a bimetallic strip) for overload protection and a magnetic mechanism (a solenoid) for short-circuit protection. The thermal mechanism responds to prolonged excess current, bending the strip to trip the breaker, while the magnetic mechanism responds instantly to high short-circuit currents. MCBs have a brief time delay to prevent unnecessary tripping during momentary surges, such as the startup of motors. They are enclosed in an insulating casing with arc chutes to dissipate arc energy and are used for circuits with voltage ratings up to 240/415 V AC and current ratings below 125 A.
  MCCB is an electromechanical circuit breaker designed for high-current applications, with ratings up to 2500 A and fault interruption capabilities ranging from 10 k to 200 k amps. It features a thermal-magnetic tripping mechanism, where the thermal (bimetal) component protects against overloading and the magnetic component responds to short circuits. The

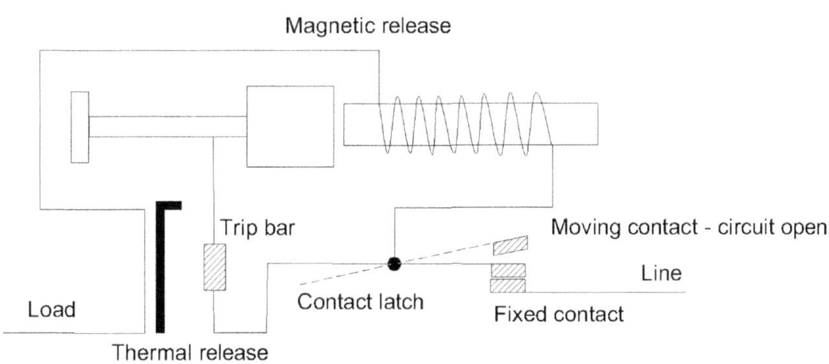


**FIGURE 5.2** MCB working principle.

operating principle is the same as the MCB philosophy given above. Unlike MCBs, MCCBs offer adjustable trip settings, making them suitable for industrial environments where normal currents exceed 100 A. MCCBs can also serve as manual switches for emergency shutdowns or maintenance, and they are available with fixed or interchangeable trip units. With fast switching times and a compact design, MCCBs are versatile, providing reliable protection in demanding electrical systems.

- *High-voltage circuit breaker*

  High-voltage circuit breakers are designed to handle extremely high fault currents in power systems operating at voltages typically above 72.5 kV, up to 800 kV or more, as shown in Fig. 5.3. Common types include vacuum circuit breakers (VCB), $SF_6$ (sulphur hexafluoride) circuit breakers, and oil circuit breakers.

  Vacuum breakers are used in medium- to high-voltage applications (up to 145 kV) and are known for their reliability and low maintenance. VCBs use vacuum as the arc extinguishing and insulating medium, with components sealed in an insulating shell. They feature a lightweight structure, small contact distance (e.g., 10 mm for 10 kV), rapid operation and compact size. The arcing time is minimal since arcs rarely form in a vacuum, typically extinguishing in half a cycle (0.01 s), making the interruption time independent of the fault current size. VCBs offer fast

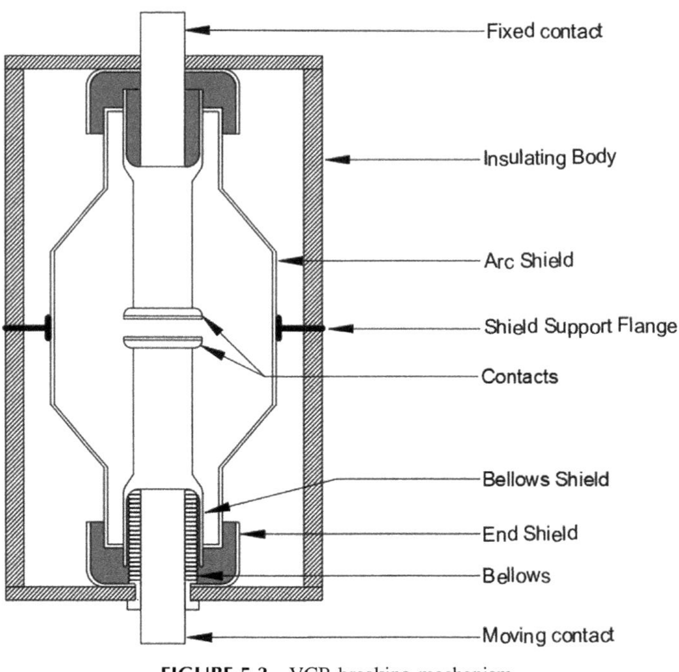


**FIGURE 5.3** VCB breaking mechanism.

recovery of the medium contact gap, long service life, low maintenance, and fireproof and explosion-proof.

$SF_6$ circuit breakers, widely used in substations and power grids, offer excellent arc-quenching capabilities and are suitable for voltages up to 800 kV, as shown in Fig. 5.4. Oil circuit breakers, although less common today, are still used in high-voltage applications, typically ranging from 72.5 to 420 kV, especially in older installations. These breakers are essential for protecting electrical grids in substations, transmission networks and industrial power systems, ensuring quick and safe interruption of fault currents to maintain grid stability and prevent equipment damage.

- *Residual current circuit breakers (RCCBs)*

  RCCBs also known as residual current devices (RCDs) are designed to protect against earth faults or leakage currents by detecting imbalances between live and neutral conductors as shown in Fig. 5.5. When a leakage current is detected, RCCBs trip to disconnect the circuit, thereby preventing electric shocks and minimizing the risk of electrical fires. They are widely used in residential, commercial and industrial settings to protect personnel and enhance electrical safety. RCCBs are typically categorized by their sensitivity to leakage currents (e.g., 30 mA for personal protection) and response time. Types include AC type RCCBs, which detect alternating currents, and A-type RCCBs, which detect both alternating and pulsating direct currents.
- *Solid state circuit breaker (SSCB)*

  SSCBs are advanced devices that use semiconductor technologies such as MOSFETs, IGBTs or SiC transistors to interrupt fault currents quickly and without the need for mechanical components as shown in Fig. 5.6. Their

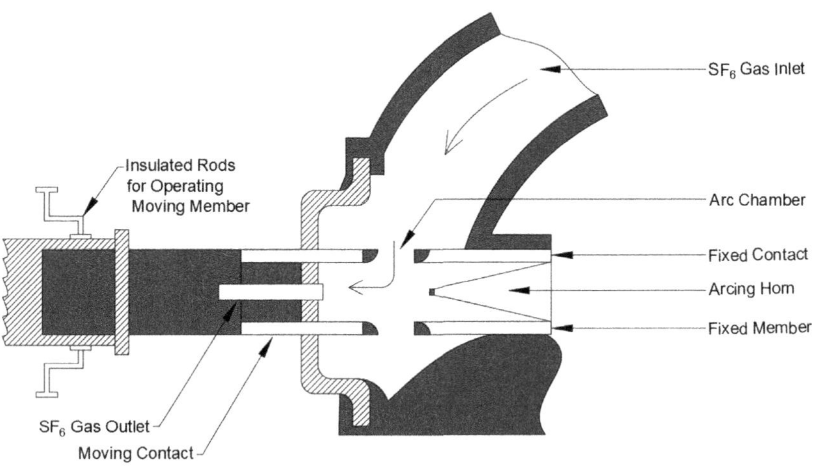


**FIGURE 5.4** $SF_6$ breaking mechanism.

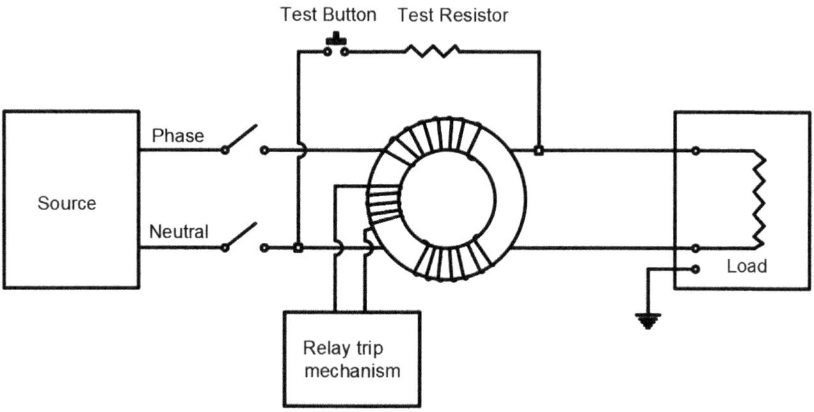


**FIGURE 5.5** RCCB methodology.

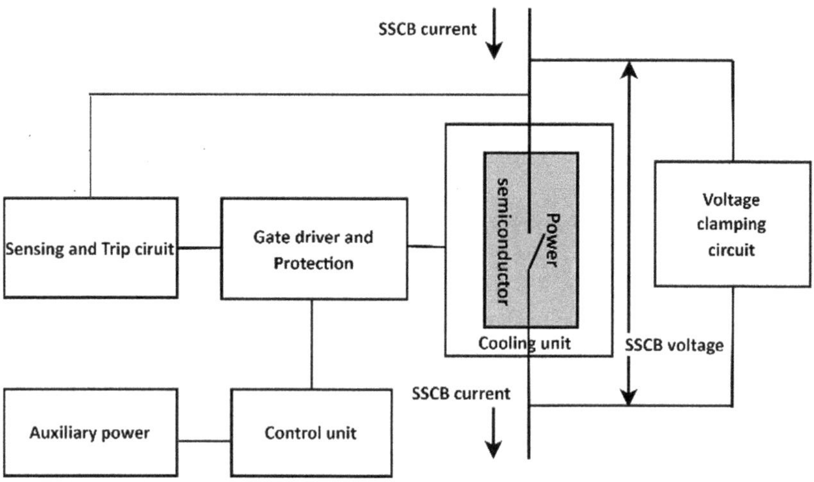


**FIGURE 5.6** SSCB component block diagram.

working principle is based on the rapid switching capability of these semiconductor devices. When abnormal conditions such as overcurrent or short circuits are detected by the control unit, the semiconductor switches are triggered to transition from a conducting to a non-conducting state almost instantaneously, typically within microseconds. This swift interruption reduces the risk of damage and eliminates the formation of electrical arcs, a common issue in traditional breakers. The control unit continuously monitors system parameters such as current and voltage using sensors and compares them against predefined thresholds, enabling real-time fault detection. SSCBs operate across a wide voltage range, from

low-voltage applications (up to 1500 VDC or 1000 VAC) to medium-voltage systems (1–52 kV). An SSCB consists of key components including a power semiconductor device, gate driver, cooling system, voltage clamping circuit, fault sensing system, sense and trip electronics, and auxiliary power supply, with the number of power semiconductors varying based on the application's voltage and current ratings, technology and breaker topology.
SSCBs offer several key advantages, including ultra-fast response times, arc-free operation and the ability to be integrated into modern power systems with programmable protection schemes. Their design includes components such as gate drivers, current and voltage sensors and cooling mechanisms to manage the heat generated by high speed switching. This makes them particularly reliable for continuous operation with minimal maintenance, as there are no moving parts that can wear out. While SSCBs are more expensive initially and require careful thermal management and EMI shielding, their rapid fault detection, longer lifespan and ability to store event logs for diagnostics enhance system reliability and efficiency. They are well suited for critical applications such as renewable energy systems, electric vehicles and high performance power grids, providing precise control over system protection.

## 2.1 Design considerations for circuit breakers

The design of circuit breakers must ensure reliable operation under various fault conditions while maintaining system stability. Circuit breakers are designed to meet the considerations mentioned in these key IEC standards – IEC 60947-1, IEC 60947-2 [1] and IEC 60898-1 [2] for low voltage systems, and IEC 62271-1, IEC 62271-100 and IEC 62271-200 [3] for high-voltage systems. Circuit breaker design must meet both functional performance and safety criteria. These design considerations are critical to ensure compliance with relevant standards while addressing various operating environments. Here, a brief technical analysis of CB design considerations that align with the requirements and specifications dictated by these standards is provided.

- *Rated voltage and current*
  The rated voltage (Ue) and rated current (In) are foundational parameters that dictate the circuit breaker's application and capacity. Circuit breakers must handle the normal operating voltage and current levels while providing protection during fault conditions. For low voltage (LV) systems, circuit breakers typically operate at voltages up to 1000 V AC or 1500 V DC. In these systems, rated currents range from 0.5 to 6300 A, and breakers must be designed to accommodate overcurrent conditions, short circuits and normal switching operations. For medium voltage (MV) and high voltage (HV) systems, circuit breakers may be rated for voltages

between 1 and 72.5 kV for MV and up to 245 kV or more for HV. Typical rated currents in these systems range from 630 to 4000 A, while HV breakers may be rated for higher currents, depending on the grid demands. The circuit breaker must reliably interrupt fault current at its rated voltage without causing insulation breakdown or excessive arcing. For low voltage systems, typical short time withstand current (Icw) ratings should allow the breaker to withstand fault conditions for at least 1 s, while; HV breakers are expected to tolerate short-time currents for 1–3 s.

- *Short circuit breaking capacity*
  Breaking capacity is one of the most critical design parameters, determining the ability of the circuit breaker to interrupt fault currents without damage. The rated short-circuit breaking capacity (Icn) of equipment is the value of short circuit breaking capacity assigned to that equipment by the manufacturer for the rated operational voltage, at rated frequency, and at a specified power factor for a.c. or time constant for d.c. systems. While designing CBs, this is usually represented as the ultimate short circuit breaking capacity (Icu) and service short circuit breaking capacity (Ics). Icu is typically represented as a breaking capacity for which the prescribed conditions according to a specified test sequence do not include the capability of the circuit breaker to carry its rated current continuously, whereas, for Ics, the circuit breaker needs to carry its rated current continuously.
  For LV circuit breakers, Icu may range from 6 to 100 kA, depending on the application, whereas, for MV and HV, Icu can be significantly higher ranging from tens of kA to several hundreds of kA. The Ics are typically tested for 25%–100% of Icu. The breaker must clear fault currents within 60 ms for LV applications and within two to three cycles (40–60 ms) for HV systems to prevent equipment damage and system instability. Another consideration is Icw, the rated value the equipment can carry without damage under test conditions. This value can range from 1 s to 3 s, for example, a 36-kV breaker may have an Icw of 25 kA for 1 s, while a 245-kV breaker must handle 63 kA for up to 3 s.
- *Arc extinction medium and technology*
  The method of extinguishing the electrical arc formed when contacts separate is a vital consideration in the design of circuit breakers. Different mediums and technologies are used depending on the system voltage level. In LV systems, air is the primary medium for arc extinction in ACBs, MCBs and MCCBs, which use arc chutes to cool and lengthen the arc, thereby increasing the dielectric strength of the air. MV and HV breakers employ more advanced technologies such as vacuum and $SF_6$ gas to extinguish arcs. Vacuum breakers, governed by IEC 62271-100, are effective for voltages up to 36 kV, providing minimal contact erosion and rapid arc quenching. $SF_6$ circuit breakers, used for voltages beyond 72.5 kV, offer high dielectric strength and rapid arc quenching, making

them suitable for systems up to 245 kV where quick fault current interruption is crucial to prevent equipment damage.
Design considerations for these breakers include ensuring rapid arc extinction and high dielectric recovery. For MV applications, vacuum interrupters must achieve arc extinction within 10 ms and maintain sufficient dielectric strength post interruption. $SF_6$ gas breakers must minimize leakage rates to ensure long-term reliability, with recovery times within 15 ms to prevent arc reignition. These design elements are critical for handling high fault currents and maintaining system stability and safety in both MV and HV systems.

- *Insulation and dielectric strength*
Insulation and dielectric strength are crucial for circuit breakers to isolate live parts, prevent electrical arcing and withstand transient overvoltage. For LV systems, IEC 60947-2 specifies insulation levels for voltages up to 1000 V AC, requiring breakers to withstand standard impulse voltages, typically around 6 kV for 400 V systems. For other systems, IEC 62271-1 defines insulation coordination for voltages up to 72.5 kV and beyond, typically, a 36-kV breaker must withstand a power frequency voltage of 70 kV for 1 min and a lightning impulse voltage of 170 kV.
Design considerations include proper allowance for clearance and creepage distance. Creepage is the shortest distance along the surface of an insulating material between two conductive parts. Clearance is the distance between two conductive parts, along a string stretched the shortest way between these conductive parts, which can be between breaker poles or between poles and the earth. These values can range from a few millimetres to several centimetres depending on the specific application, pollution level and environmental and insulation requirements. For example, according to IEC 62271-1, creepage distances can be a minimum of 900 mm for a 36 kV system in high pollution conditions. For MV and HV systems, insulating materials must endure both electrical and thermal stress, with technologies such as $SF_6$ gas breakers requiring minimal gas leakage and rapid dielectric recovery to prevent arc reignition. These measures ensure the reliable operation of circuit breakers under various electrical and environmental stresses.
- *Tripping mechanism*
Circuit breakers are designed with different tripping mechanisms, primarily thermal-magnetic and electronic. Thermal magnetic circuit breakers are widely used in low- and medium-voltage applications. The thermal component, based on a bimetallic strip, responds to overload conditions by bending and tripping the breaker when the current exceeds the rated value for an extended period, typically between 2 s and 2 min. The magnetic component trips instantaneously in short-circuit conditions, typically responding within milliseconds when the current exceeds 5–10 times the rated value. Electronic circuit breakers, on the other hand, offer

programmable settings and more precise protection. They are often used in industrial applications for their flexibility in adjusting trip parameters and their ability to provide ground fault protection and remote monitoring.

Selecting the correct trip curve is vital to the protection of electrical systems. Circuit breakers have specific time–current characteristic curves, such as type B, C and D, which define their response to overcurrent conditions. Type B breakers trip at 3–5 times the rated current and are typically used for resistive loads such as lighting circuits. Type C breakers trip at 5–10 times the rated current and are suitable for inductive loads such as small motors and transformers, while type D breakers trip at 10–20 times the rated current, making them ideal for heavy machinery with high inrush currents. Standard references such as IEC 60947-2 provide guidance on these trip characteristics, ensuring they are appropriately matched to the application's load type to prevent nuisance tripping or inadequate protection.

When designing a circuit breaker, the selection of the tripping mechanism and the trip curve should consider the nature of the load, starting currents and fault current levels. Thermal-magnetic breakers are often used for simpler applications, while electronic breakers are preferred in industrial settings requiring high adjustability and real-time diagnostics. Standards such as IEC 60947-1 and IEC 60947-2 guide the design process for low-voltage circuit breakers, ensuring proper coordination between the breaker's trip settings and the system's protection needs. For medium- and high-voltage applications, IEC 62271-1 and IEC 62271-100 govern the technical requirements, ensuring that circuit breakers can withstand fault currents and arc interruptions while offering adjustable trip characteristics to suit various operating conditions.

- *Operating mechanism*

  The operating mechanism of circuit breakers is a fundamental design consideration that directly impacts their performance, reliability and suitability for various applications. Circuit breakers can operate either manually or automatically, with manual operation commonly seen in MCBs and MCCBs, where the user manually resets the breaker after it trips. This manual reset feature allows for greater operator control in applications where it is critical to assess fault conditions before re-energizing the circuit. However, for more complex and safety sensitive applications, automatic tripping mechanisms are essential. These mechanisms ensure that circuit breakers can detect fault conditions such as overcurrent or short circuits and respond instantly to isolate the affected circuit without human intervention, as specified in IEC 60947-2.

  The choice of operating mechanism can vary significantly based on the application requirements. Spring operated mechanisms are commonly used in low-voltage circuit breakers. In these designs, a mechanically charged spring stores energy, which is released to trigger the contact separation during fault conditions. This rapid release of energy allows for quick operation, making it ideal for applications where prompt fault isolation is necessary. In contrast, magnetic operated circuit breakers employ an electromagnetic coil that activates the mechanism to trip the breaker when fault conditions are detected. This design is prevalent in circuit breakers designed for higher fault current scenarios, where the speed of operation is critical to prevent equipment damage and maintain system integrity.

For high-voltage circuit breakers, more robust operating mechanisms are required to handle greater mechanical forces and energy levels. Hydraulic or pneumatic operation is often utilized in these applications, leveraging fluid pressure or air to actuate the tripping mechanism. This type of operation allows for greater force and reliability when opening or closing contacts under load. Standards such as IEC 62271-100 specify the performance requirements for high-voltage circuit breakers, including operational mechanisms, ensuring that they can effectively handle the stresses associated with high-voltage conditions.

- *Temperature rise and heat dissipation*

  Temperature rise is a critical design parameter for circuit breakers, as excessive heat can damage components or reduce their operational life. For LV breakers, the temperature rise for copper contacts should not exceed 60°C above ambient temperature, while insulating materials should not rise more than 40°C above ambient. For other breakers, the temperature rise for current-carrying parts, such as terminals, should not exceed 70°C above ambient.

  Design considerations include incorporating adequate heat dissipation measures, such as air cooling or ventilation, especially for breakers operating at higher continuous currents. For HV breakers using $SF_6$ gas, temperature management must ensure the gas's thermal stability and consistent dielectric properties under various fault conditions. Proper ventilation and cooling mechanisms, like heat sinks or forced air cooling, are essential to handle the heat generated during normal and fault conditions, preventing insulation breakdown and extending the operational life of the circuit breakers.

- *Mechanical operation and endurance*

  The mechanical operation of circuit breakers must be robust and reliable to ensure long operational life with minimal wear and tear, even under frequent switching conditions. IEC 60947-1 specifies a mechanical endurance of at least 8500 operations at rated current. For medium and high voltage applications, IEC 62271-100, require a minimum of 2000 mechanical operations, with operating mechanisms providing closing and opening times within two to five cycles.

  Design considerations include the use of spring-operated or motor-driven mechanisms for consistent performance, even during emergencies or fault scenarios. Additionally, circuit breakers in seismic zones must have enhanced mechanical robustness to withstand seismic forces without malfunction, as outlined in IEC 62271-1. Proper design ensures high mechanical reliability and long term operation, particularly in HV systems where interruptions are less frequent but require absolute dependability.

- *Protection coordination and selectivity*
  Protection coordination is essential to ensure that only the circuit breaker closest to a fault trips, maintaining the rest of the system's operation. IEC 60947-2 mandates that LV breakers support selective coordination with upstream and downstream protective devices, based on current–time curves. For high-voltage breakers, IEC 62271-100 requires selective tripping to minimize disruption. Design considerations include configurable trip units for accurate overload and short circuit protection settings, with electronic trip units preferred for HV applications to enhance fault-clearing precision and selectivity. This coordination ensures system reliability and minimizes damage to critical infrastructure.
- *Environmental considerations*
  Circuit breakers must be designed to withstand a variety of harsh environmental conditions, including temperature extremes, humidity and pollution. IEC 62271-100 specifies environmental tests for high voltage (HV) breakers to ensure reliable performance under such conditions. For installations in seismic regions, IEC 62271-1 mandates that circuit breakers be designed to withstand seismic shocks, based on expected severity, typically following specific g-levels and frequencies. Additionally, environmental factors such as humidity, temperature variation and corrosive atmospheres must be accounted for, particularly for outdoor HV breakers. The use of $SF_6$ gas in sealed designs helps prevent moisture ingress and corrosion, ensuring long term reliability. Standards such as IEC 60068-2 provide additional guidelines for environmental testing, including temperature and humidity cycling, to ensure that breakers can endure and operate effectively in diverse conditions.
  At higher altitudes, the dielectric strength of air is reduced, necessitating increased creepage distances and clearances to maintain insulation integrity. For instance, at an altitude of 1000 m, the dielectric strength drops by approximately 10%, requiring adjustments to the insulation design. Circuit breakers must also be capable of operating reliably under extreme temperatures, typically ranging from −30°C to +40°C for HV systems. Compliance with standards such as IEC 62271-304, which covers the performance of HV equipment under severe environmental conditions, is crucial. These standards ensure that circuit breakers can withstand environmental stresses, maintain safety and provide reliable operation across various applications and geographic locations.

## 3. Relay fundamentals

Relays are electrically operated switches that control the opening and closing of a secondary circuit based on signals received from the primary circuit, electromechanically or electronically. A relay can exist in two states: normally open (NO), indicating an open contact when the relay is not energized, or a normally closed (NC), indicating a closed contact when the relay is not energized. In either case, applying electrical current to the contacts will change their state. Relays can be classified into different categories based on their properties and specific applications. Based on operating principles, relays can be classified as follows:

- *Electromechanical relay (EMR)*
  Electromechanical relays (EMRs), as shown in Fig. 5.7, operate using an electromagnetic coil that generates a magnetic field when energized by a control signal. This field pulls a movable armature, causing the relay's contacts to either open or close, depending on the configuration (e.g., SPST, DPST). The relay consists of key components such as the coil, armature, contacts and a return spring that repositions the armature when the coil is de-energized. EMRs are used with both AC and DC sources, although DC coils typically include a freewheeling diode to protect against back EMF. While effective, a major drawback of EMRs is the arcing that

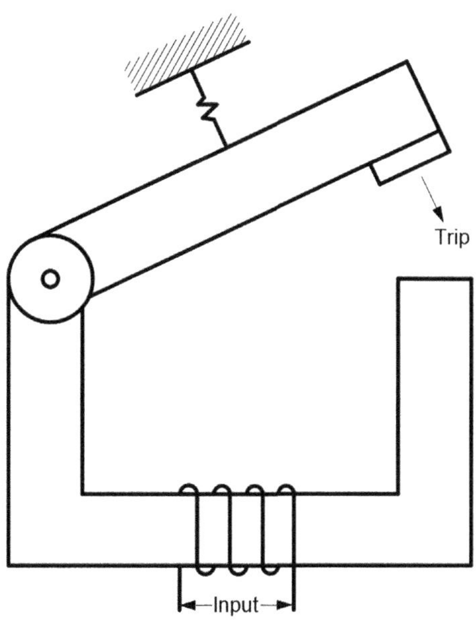


**FIGURE 5.7** EMR with movable armature.

occurs during contact switching, which increases resistance and reduces the relay's lifespan over time.

- *Solid-state relay (SSR)*
SSRs, as shown in Fig. 5.8, use semiconductors instead of mechanical components, enabling faster, noiseless and more durable switching. They work on the principle of opto-isolation, where an LED activates a photo-detector, triggering a semiconductor device (such as a thyristor, TRIAC or MOSFET) to switch the load circuit without physical movement. SSRs consist of an input circuit (activated by control voltage), a control circuit (coupling input and output) and an output circuit (which switches the load). While SSRs offer high speed operation, low power consumption and a longer lifespan, their main disadvantage is the heat generated by voltage drop across the semiconductors.
- *Hybrid relay*
Hybrid relays combine the advantages of both SSRs and EMRs to overcome their individual drawbacks. In a hybrid relay, the SSR and EMR operate in parallel as shown in Fig. 5.9. Initially, the SSR switches on and takes the load current, eliminating contact arcing in the EMR. Once the EMR is fully engaged, the SSR is deactivated, and the EMR carries the entire load, preventing heat loss from the SSR. This design reduces both power loss due to heat and arcing, offering efficient and reliable switching.
- *Computer relay*
Computer relays, also known as microprocessor, digital or numerical relays, are advanced protection devices used in electrical systems to detect faults, ensure system stability and improve overall reliability as shown in Fig. 5.10. These relays process signals from current and voltage

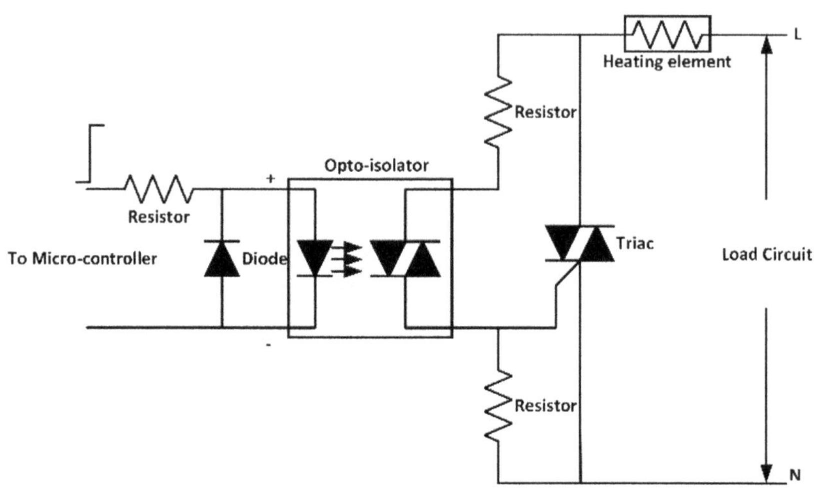


**FIGURE 5.8** SSR operating principle.

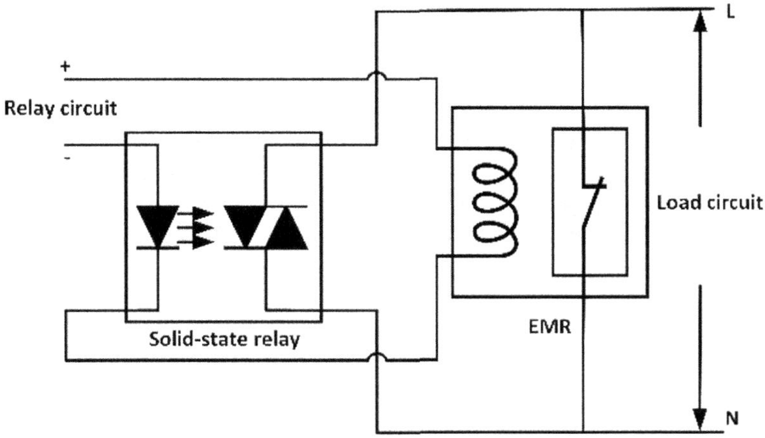


**FIGURE 5.9** Hybrid relay principle.

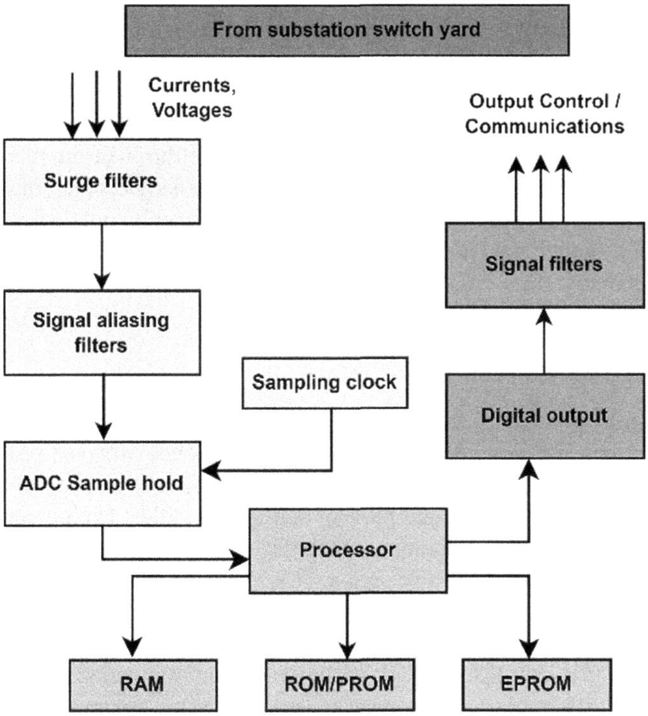


**FIGURE 5.10** Computer relay block diagram.

transformers using embedded microprocessors and sophisticated algorithms to analyze system conditions and trigger appropriate protective actions. They offer greater precision, flexibility and speed compared with traditional electromechanical or static relays, as they can implement multiple protection functions – such as overcurrent, differential and distance protection – within a single device. With programmable settings, computer relays allow users to customize protection schemes based on specific requirements. Additionally, these relays support real-time monitoring, diagnostics and communication through modern protocols such as IEC 61850 [4], making them essential for smart grids and substation automation systems.

Computer relays function by converting analog signals from the power system, such as voltages and currents, into digital form through surge and anti aliasing filters. The signals are sampled at high frequencies (up to 64 to 128 times the system's fundamental frequency) using analog to digital converters (A/D) for accurate digital representation. Microprocessors then execute various relaying algorithms to detect faults by comparing the digital data against predefined thresholds. These relays offer self diagnostics, can adapt to changing system conditions and facilitate remote monitoring through communication interfaces. Their compact design, ease of configuration and ability to store event logs and waveform data improve fault analysis and reduce downtime, contributing to the efficiency and robustness of modern electrical protection systems.

The number of switches inside a relay is called the poles, and the number of circuits being controlled per pole is called the throw of the relay. Based on poles and throw, relays can be classified as single pole single throw (SPST), single pole double throw (SPDT), double pole single throw (DPST), double pole double throw (DPDT), etc. A relay can have as many as 12 poles.

## 3.1 Design considerations for relays

In electrical power applications, relay design is crucial for ensuring reliable protection, coordination and overall system performance. Relays must respond accurately to abnormal conditions such as overloads, short circuits and voltage fluctuations to isolate faults and prevent damage to critical equipment. Design considerations for relays encompass technical, operational and environmental factors, including the relay type, sensitivity, trip settings, communication capabilities and environmental resilience. Additionally, adherence to international standards such as IEC 60255 (measuring relays and protection equipment) [5] and IEC 61850 (communication networks for substation automation) [4] is critical for ensuring compatibility, reliability and safety across various electrical systems. Standards such as IEEE C37 [6] guide relay performance under electrical, mechanical and environmental conditions, further reinforcing relay reliability in power system protection. By following these standards and focussing on proper coordination with other protection devices, relays can effectively protect electrical systems, minimizing downtime and ensuring safe operation. A brief analysis of various design considerations is provided in the following.

- *Relay-type selection*
  Selecting the appropriate relay type is fundamental in the design of protection systems, ensuring that the relay matches the fault scenarios and operating conditions of the network. Overcurrent relays, governed by IEC 60255-151, are widely used in transmission and distribution systems to protect against short circuits and overloads. These relays operate when the current exceeds a predefined value, with time—current characteristic curves ensuring proper coordination with other devices. Distance relays, standardized under IEC 60255-121, provide protection for transmission lines by detecting faults based on impedance measurement between the relay and the fault point. They are crucial for high-voltage networks, where fast and reliable detection of distance faults can prevent widespread outages. Differential relays, often applied in transformer, generator and busbar protection, detect discrepancies between the current entering and leaving the protected zone. According to IEC 60255-127, differential relays can rapidly identify internal faults with high accuracy, minimizing damage to critical assets. Directional relays, also described in IEC 60255, are essential in systems with multiple power sources, as they determine the direction of fault current flow. This is critical for coordinating protection in complex grids such as interconnected networks or systems with distributed generation. The choice of relay type must be based on system configuration, fault conditions and the need for coordination with other protection devices.
- *Operating time and speed*
  The operating time of relays is a key consideration in ensuring the quick isolation of faults, thereby minimizing equipment damage and reducing system downtime. Relays must operate fast enough to clear faults but also be coordinated with other protection devices such as circuit breakers to avoid unnecessary tripping. IEC 60255-3 specifies performance characteristics for protection relays, including their operating time and accuracy. For example, instantaneous relays respond within milliseconds, typically operating without intentional time delay to quickly isolate severe faults such as short circuits. These are crucial in applications requiring fast fault clearance, such as protecting transformers or busbars, where delays could cause significant damage.
  In contrast, time delay relays are designed with adjustable delay settings to ensure coordination with upstream and downstream devices, allowing selective tripping. According to IEC 60255-151, [5] these relays use inverse or definite time—current characteristics to provide graded protection, where devices farther from the fault trip first. The typical response times

for time-delay relays can range from a few milliseconds to several seconds, depending on the fault condition and coordination requirements. For example, inverse-time overcurrent relays use a time–current curve that allows faster tripping for higher fault currents, ensuring a balance between speed and selectivity. Following these time delay principles ensures that relays provide precise, graded fault isolation in complex networks while maintaining coordination with circuit breakers and avoiding system-wide outages.

- *Sensitivity and pickup settings*
  The sensitivity of a relay determines its ability to detect low-magnitude faults, which is essential for effective protection and coordination across the power system. IEC 60255-151 outlines guidelines for setting relay sensitivity and pickup current to ensure accurate detection while preventing false trips. The pickup current is the minimum current level at which the relay operates, and it must be carefully set to balance protection and system stability. For overload protection, the pickup current is typically set between 1.2 and 1.5 times the full load current, ensuring that normal operating conditions, such as load fluctuations, do not trigger unnecessary tripping. This allows the relay to detect sustained overloads that could damage equipment, but without reacting to temporary overcurrent.
  For short-circuit protection, the pickup setting is much higher, usually between 6 and 12 times the full load current, allowing the relay to respond quickly to severe faults while maintaining immunity to transient conditions such as motor inrush currents. The accuracy of these settings is crucial for avoiding relay misoperation and ensuring coordination with other protective devices. IEC 60255-3 specifies the performance requirements for the accuracy and repeatability of pickup thresholds to ensure that the relay reliably detects faults under all operating conditions. Properly configured pickup settings also contribute to selective coordination, where the relay closest to the fault operates first, avoiding unnecessary trips in other parts of the system.
- *Power supply requirements*
  Power supply requirements are critical for the reliable operation of relays in electrical power applications, ensuring they function effectively under various operational conditions. Relays must be designed to operate with the specified control voltage, whether it is AC or DC, as dictated by system design. IEC 60255-6 outlines the performance requirements for auxiliary power supplies, specifying that relays should operate correctly within their rated voltage ranges, which typically include control voltages of 24 V, 48 V, 110 V or 220 V, depending on the application.
  Moreover, the voltage tolerance is crucial; relays should be capable of functioning correctly under voltage variations of approximately ±10% of the nominal rated voltage. This tolerance ensures that relays can withstand common fluctuations in the power supply, such as those caused by load

changes or equipment operation. Additionally, IEC 60255-21 specifies requirements for the dielectric strength and insulation resistance of relays, which are essential for ensuring safe operation under varying voltage conditions.

The power supply's stability is vital not only for the relay's functionality but also for ensuring that protection schemes operate effectively during fault conditions. For instance, relays powered by DC supplies may have different response characteristics compared with those powered by AC, necessitating careful consideration during design and selection.

- *Coordination with circuit breakers*

  Effective coordination between relays and circuit breakers is essential for ensuring that the nearest protection device trips first during a fault condition, thereby minimizing the impact on the overall power system. Proper coordination is achieved by carefully matching the time–current characteristic curves of relays and circuit breakers, which delineate how quickly each device will respond to varying levels of fault current. Standards such as IEC 60255-151 provide guidance on establishing these curves to ensure selective coordination, which is crucial for preventing unnecessary outages and maintaining system stability. For instance, when a fault occurs, the relay closest to the fault should operate first, allowing downstream circuit breakers to remain operational and continue supplying power to unaffected areas.

  In applications involving distance relays, zone protection is a key concept that enhances coordination across different sections of the system. According to IEC 60255-121, distance relays typically operate in multiple zones (e.g., zone 1, zone 2, zone 3), each with distinct settings and time delays. Zone 1 provides the fastest response, usually covering 80%–90% of the protected line's length, ensuring immediate disconnection of faults close to the relay. Zone 2 typically covers the remainder of the line with a slight time delay, allowing for coordination with other protective devices further upstream. Finally, Zone 3 serves as backup protection, with an even longer delay to account for coordination with relays and breakers on adjacent lines. By implementing such zone-based protection strategies, designers can ensure a graded response across the system, enhancing reliability and reducing the risk of cascading failures during fault conditions.

- *Communication protocols*

  In the era of smart grids, communication protocols have become essential for modern relays to enable seamless integration with various power system components, including supervisory control and data acquisition (SCADA) systems and remote monitoring applications. The IEC 61850 standard is pivotal in this regard, defining a comprehensive framework for communication in substation automation, allowing for interoperability between protection relays, circuit breakers and other devices within the power system. This standard facilitates the exchange of real-time data, enabling operators to monitor system performance, detect faults and implement coordinated control measures effectively.

The IEC 61850 architecture is based on an object-oriented approach, where devices communicate using standardized data models and services. For example, GOOSE (Generic Object-Oriented Substation Event) messaging allows for high-speed communication of time-critical events, such as fault detection, enabling rapid relay operation and reducing coordination times. This is particularly beneficial in scenarios where fast disconnection of a fault is essential, such as in high-voltage transmission lines, where delays can lead to widespread outages.

Additionally, MMS (Manufacturing Message Specification) facilitates the exchange of diagnostic information and configuration data, allowing for remote management and maintenance of relays. By employing these communication protocols, power system operators can achieve enhanced situational awareness and optimize protective relay settings based on real-time operational data.

Moreover, the interoperability offered by IEC 61850 ensures that various devices from different manufacturers can work together seamlessly, promoting flexibility and scalability in power system design. This is critical in complex scenarios involving distributed energy resources (DERs) and microgrid configurations, where multiple generation sources and storage devices must coordinate effectively to maintain system stability. By leveraging advanced communication protocols, modern relays can play a crucial role in enhancing the reliability and resilience of electrical power systems.

- *Reliability and redundancy*

  Ensuring reliability and redundancy in relay design is essential for maintaining the protection of electrical systems, particularly in critical applications such as transformers and generators. Backup relays serve as an essential component of a robust protection scheme, providing fail-safe mechanisms if the primary relay fails to operate. Standards such as IEC 60255-27 outline the need for redundancy in protective devices, emphasizing that multiple relays may be employed to ensure continuous protection, thereby mitigating the risk of system outages caused by a single point of failure.

  In addition to backup relays, the implementation of redundancy strategies is critical in applications where reliability is paramount. For instance, dual-redundant systems may be configured to allow one relay to take over the protection function if the other becomes inoperative, ensuring that the system always remains protected. This approach is particularly important in environments where the consequences of failure could lead to significant financial loss or safety hazards.

- *Calibration and fail-safe design*
Regular testing and calibration of relays are necessary to verify trip accuracy and the integrity of time−current characteristics over time. IEC 60255-11 provides guidelines for routine testing of protection relays, highlighting the importance of conducting functional tests and verifying that the relay operates within its specified parameters. Calibration processes may involve adjusting the relay settings to ensure that they accurately reflect the required pickup values and time delays, which are crucial for effective fault detection and system coordination.
Incorporating a fail-safe design into relay construction is another key consideration. Relays should be engineered to fail in a mode that prevents exacerbating system failures, ensuring that any malfunction does not lead to cascading outages. According to IEC 60255-26, fail-safe mechanisms may involve settings that allow the relay to default to a non-operational state or to signal an alarm condition, alerting operators to the need for intervention. Also, modern relays often feature self-monitoring and diagnostic capabilities, enabling them to perform internal checks and identify faults in real time. These functionalities allow relays to alert operators to any internal issues, enhancing system reliability and facilitating timely maintenance
- *Environmental considerations*
Environmental considerations play a critical role in the design and selection of relays for electrical power applications, ensuring their reliable operation under varying conditions. Temperature range is a significant factor; relays should be designed to function effectively within the temperature extremes anticipated in their operating environment. According to IEC 60255-1, relays must be rated for specific temperature ranges, typically from −40°C to +55°C, to accommodate different climatic conditions. This temperature range ensures that relays can perform consistently without risk of failure due to overheating or freezing, which is especially important in regions with extreme weather.
Additionally, relays operating in high-power environments are susceptible to electromagnetic interference (EMI), which can lead to erratic behaviour or malfunction. To mitigate this risk, relays should incorporate shielding and filtering techniques that comply with IEC 60255-26, which outlines requirements for immunity to electromagnetic disturbances. For example, using metal enclosures or integrated EMI filters can significantly enhance a relay's resilience against external noise, ensuring accurate operation in environments such as substations or industrial plants where high-frequency noise is prevalent.
Another important environmental consideration is the Ingress Protection (IP) rating of the relay, particularly for applications exposed to outdoor conditions or harsh environments. Relays must be rated appropriately − commonly IP54 or higher − to protect against dust, moisture and other

contaminants, as specified in IEC 60529. An IP54 rating indicates that the relay is protected against limited dust ingress and water splashes from any direction. This is crucial for applications such as renewable energy installations, where equipment is exposed to outdoor elements and requires robust protection to ensure reliable operation and longevity.

- *Physical size and mounting*
  The physical size and mounting options for relays are crucial considerations in the design and installation process, ensuring that protective devices can be effectively integrated into existing infrastructure such as switchgear, control panels or substations. As outlined in IEC 60255-22, relays must adhere to specific dimensional and weight requirements that facilitate compatibility with standard mounting systems. The space available for relay installation can significantly impact the choice of relay, particularly in compact switchgear where space constraints are a primary concern.
  Mounting options can vary widely based on the application and environment. For instance, flush-mounted relays are often utilized in control panels where aesthetics and space-saving designs are prioritized. This type of mounting allows for a sleek integration into the panel's faceplate, ensuring easy access while maintaining a compact footprint. In contrast, rack-mounted relays are suitable for larger systems where equipment may need to be arranged in a modular format for easy accessibility and maintenance. This type of installation is commonly seen in substation control rooms where multiple relays are housed within a standard equipment rack, facilitating organized management of protective devices.
  DIN rail-mounted relays offer a flexible and convenient mounting solution, allowing for easy installation and removal in various industrial applications. DIN rail systems are widely adopted due to their adaptability, enabling quick modifications or expansions of control systems without extensive rewiring. According to IEC 60715, DIN rail specifications ensure that relays can be securely fastened, providing reliable performance even in vibration-prone environments.
- *Cybersecurity*
  Cybersecurity is a fundamental aspect of modern relay design, as outlined in standards such as IEEE C37.240 and IEEE 2030.100 [7], ensuring the protection and resilience of digital relays in interconnected and smart grid environments. Computer relays, which are integral to modern power systems, rely on real-time communication with other devices in substations and across wide-area networks, making them potential targets for cyberattacks. IEEE C37.240 [6] establishes guidelines for securing communications and data exchanges within protection and control systems, emphasizing the importance of data integrity, confidentiality, authentication and access control. Measures such as the implementation of

encryption protocols, secure access points, role-based access controls and logging for detecting unauthorized access attempts are suggested by the standard.

Moreover, IEEE 2030.100 provides a comprehensive framework for grid cybersecurity, extending to digital protection systems such as relays. This standard emphasizes the importance of a defence-in-depth strategy, incorporating multiple layers of protection such as intrusion detection systems (IDS), firewalls and secure firmware updates. Relays must now be equipped to handle encrypted data exchanges and are designed with secure boot processes, ensuring that malicious software cannot infiltrate the system. As relays communicate over Ethernet-based protocols such as IEC 61850, these cybersecurity mechanisms ensure real-time data integrity and minimize the risk of external manipulation or malicious commands.

## 4. Conclusion

In conclusion, the design considerations for circuit breakers and relays play a pivotal role in the reliability, safety and efficiency of electrical power systems. This chapter has discussed various critical aspects of these devices, including relay type selection, operating mechanisms, sensitivity settings and environmental considerations, emphasizing the need for careful integration of these elements into the overall design framework.

The selection of appropriate relay types – such as overcurrent, distance and differential relays – ensures that protection schemes are tailored to the specific requirements of the system. Understanding the operating mechanisms of circuit breakers, whether manual or automatic, and the implications of spring-operated, magnetic-operated or hydraulic systems, further enhances their effectiveness in fault isolation and system protection.

Moreover, the importance of adherence to international standards, such as IEC 60255, IEC 60947 and IEC 62271, cannot be overstated. These standards provide essential guidelines for the performance, testing and operational integrity of relays and circuit breakers, ensuring that they meet the rigorous demands of modern electrical applications. Regular testing and calibration, combined with considerations for reliability and redundancy, fortify the operational resilience of these protective devices.

Environmental factors, including temperature range, electromagnetic interference and ingress protection, must also be addressed to ensure reliable operation under varying conditions. The careful consideration of physical size and mounting options facilitates seamless integration into existing systems, enhancing maintenance and accessibility.

In summary, the design of circuit breakers and relays is a multifaceted process that requires a comprehensive understanding of technical requirements, operational contexts and environmental challenges. By prioritizing these considerations, engineers can develop robust protection schemes that safeguard electrical systems, mitigate risks and ensure sustained operational performance in an increasingly complex energy landscape. The insights gained from this chapter will serve as a foundation for advancing the design and application of these vital components in future electrical infrastructure projects.

## Acknowledgements

The authors would like to acknowledge Ms. Deepa Subedi from the Real-time Power Systems Laboratory at the University of Saskatchewan for her help in drawing some of the figures.

## References

[1] IEC 60947 Series—Low-Voltage, Switchgear and Controlgear (IEC 60947—1—Part, IEC 60947–2.

[2] IEC 60898 Series—Electrical Accessories—Circuit-Breakers for Overcurrent Protection for Household and Similar Installations (IEC 60898—1, IEC 60898—2, IEC 60898—3).

[3] IEC 62271 Series—High-Voltage Switchgear and Controlgear (IEC 62271—1, IEC 62271—100, IEC 62271—200).

[4] IEC 61850 Series—Communication Networks and Systems for Power Utility Automation (IEC 61850—90-1, IEC 61850—9-2, IEC 61850—7-2).

[5] IEC 60255 Series—Measuring Relays and Protection Equipment (IEC 60255—1, IEC 60255—3, IEC 60255—151, IEC 60255—121, IEC 60255—127).

[6] IEEE C37 Series—IEEE Standard for Circuit Breakers and Relays (IEEE C37.06, IEEE C37.90, IEEE C37.99, IEEE C37.100, IEEE C37.240).

[7] IEEE 2030.100—IEEE Recommended Practice for Implementing an IEC 61850-based Substation Communications, Protection, Monitoring, and Control System.

# Part II

# Control and instrumentation and self-sustaining ecosystems

Chapter 6

# Control and instrumentation aspects of switchgear operation

Sovan Das[1] and Soumya Debashis Das[2]
[1]*Kyocera SLD Laser Inc, Goleta, CA, United States;* [2]*Veer Surendra Sai University of Technology, Department of Electrical & Electronics Engineering, Burla, Odisha, India*

## 1. Introduction

Switchgear plays a critical role in electrical power systems by controlling, protecting and isolating electrical equipment. It forms a crucial link between the power source and the end user, ensuring the safe and efficient distribution of electricity. Control and instrumentation aspects are integral components of switchgear, providing the necessary tools to monitor, regulate and safeguard the electrical network. In this introduction, we delve into the fundamental concepts of control and instrumentation in switchgear, highlighting their significance in ensuring the reliability and performance of power systems.

### 1.1 Control in switchgear

Control systems in switchgear are designed to manage the operation of electrical equipment, allowing operators to manipulate the configuration of the network as per the demand and operational requirements. The primary goal is to ensure the seamless and safe flow of electrical power. Modern switchgear incorporates advanced control technologies, such as programmable logic controllers (PLCs) and intelligent electronic devices (IEDs), to enable precise monitoring and control of the equipment.

The control aspect involves various functionalities, including the following:

1. Circuit breaker operation: Switchgear employs circuit breakers to interrupt the flow of current during abnormal conditions or faults. Control systems

**Switchgear Design, Operation, and Maintenance Using Industry Standards**
**https://doi.org/10.1016/B978-0-443-24718-7.00006-8**

enable the automatic or manual operation of circuit breakers, ensuring a rapid response to faults and minimizing downtime.

2. Load management: With the increasing complexity of power systems, effective load management becomes crucial. Control systems in switchgear facilitate the seamless transfer of loads, ensuring optimal utilization of resources and preventing overloads.
3. Fault detection and diagnosis: Control systems incorporate sensors and monitoring devices to detect faults or abnormal conditions within the electrical network. Rapid fault detection allows for quick isolation and restoration of the system, reducing the impact of failures.
4. Communication interfaces: In modern power systems, effective communication between switchgear components is vital. Control systems integrate communication interfaces, enabling seamless data exchange and coordination between various elements of the power distribution network.

## 1.2 Instrumentation in switchgear

Instrumentation refers to the use of various devices and sensors to measure, monitor and control the electrical parameters within the switchgear.

Accurate instrumentation is crucial for maintaining the integrity and efficiency of the power system as shown in Fig. 6.1 [1]. Key aspects of instrumentation in switchgear include the following:

1. Voltage and current monitoring: Instrumentation devices measure the voltage and current levels at different points in the electrical network. This information is essential for assessing the system's health, identifying abnormalities and making informed decisions regarding load distribution.

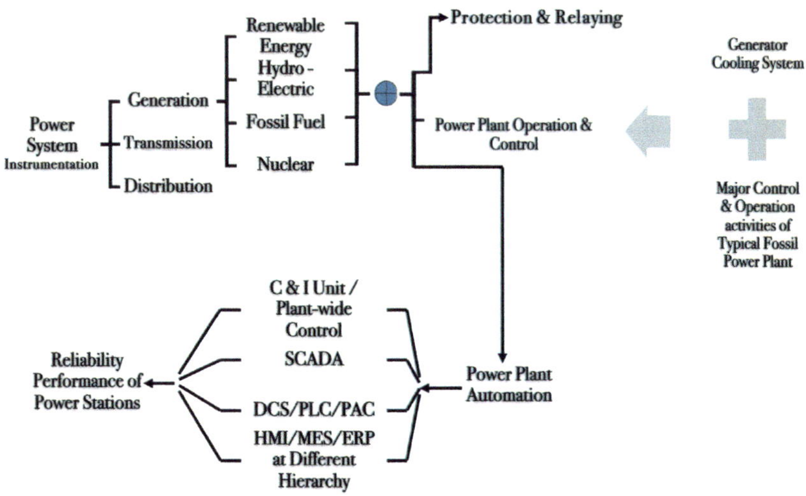


FIGURE 6.1 Hierarchy of power system instrumentation.

2. Temperature and humidity sensors: Switchgear operates in various environmental conditions. Monitoring temperature and humidity ensures that the equipment operates within specified limits, preventing overheating and damage to sensitive components.
3. Gas monitoring in gas-insulated switchgear (GIS): GIS involves the use of insulating gas for compact design. Instrumentation in GIS includes gas sensors to monitor the condition of the insulating gas, enabling timely detection of leaks or degradation.
4. Protection relays: Instrumentation devices include protection relays that act as the first line of defence against faults. These relays respond to abnormal conditions by initiating protective actions, such as tripping circuit breakers, to safeguard the equipment and the overall power system.

In conclusion, the control and instrumentation aspects of switchgear are essential for the reliable and efficient operation of electrical power systems. The integration of advanced technologies in control systems and instrumentation devices enhances the capability to monitor, regulate and protect the network, ensuring the delivery of electricity safely and uninterruptedly. As power systems continue to evolve, the role of control and instrumentation in switchgear becomes increasingly critical in adapting to the dynamic demands of modern electrical grids.

## 2. Need for control and instrumentation in the industry environment

Control and instrumentation (C&I) serves as a crucial interdisciplinary domain vital for the functionality and performance of industrial facilities. C&I engineers play a pivotal role in crafting, installing, and upkeeping the control systems and accompanying instrumentation utilized for operating and/or automating industrial procedures. Proficiency in various fields such as chemistry, mechanics, electricity, magnetism, electronics, microcontrollers, microprocessors, software languages, process control, pneumatics, hydraulics and communications is imperative in this field. Essentially, C&I acts as the nervous system for industrial complexes, power generation and any processes requiring intelligence to achieve the task of producing a product or managing a process.

Fundamental concepts and principles governing industrial plant operations encompass measurements of flow, level, temperature and pressure, as well as electronics and pneumatics instrumentation, control loops and PID control, among others [2]. A measurement instrument serves as a device capable of detecting changes, whether physical or otherwise, in a specific process and translates these changes into understandable information for the user.

The significance of C&I in the industrial milieu cannot be emphasized enough. C&I systems are deployed to oversee and regulate diverse processes in the production of goods and services. They play a pivotal role in ensuring the efficiency and effectiveness of processes and adherence to required quality standards. Additionally, C&I systems contribute to worker and environmental safety by detecting and preventing accidents and mitigating their impact when they do occur [3].

A notable advantage of C&I systems lies in their ability to diminish the reliance on manual labour in industrial processes. Through the automation of numerous production processes, these systems mitigate the risk of human error, enhance production efficiency and subsequently reduce costs while increasing profits for the companies involved. Furthermore, C&I systems contribute to lessening the environmental impact of industrial processes by monitoring and controlling various production facets, ultimately reducing waste generation and energy consumption [4]. This eco-friendly approach helps diminish the carbon footprint of industrial operations and promotes sustainable development.

In conclusion, C&I stands as a critical facet of the industrial landscape, ensuring the seamless and effective operation of industrial plants. C&I systems are indispensable for monitoring and regulating processes in the production of goods and services, prioritizing worker safety and minimizing environmental impact [5]. By automating processes, these systems reduce manual labour, enhance efficiency and contribute to sustainable industrial practices. The field of C&I demands a broad spectrum of knowledge and skills, making it an essential component for the optimal functioning of industrial facilities.

In the contemporary industrial landscape, characterized by rapid technological advancements and increasing complexity, the role of C&I has become indispensable. The integration of sophisticated control systems and precise instrumentation is critical for ensuring the efficiency, safety and competitiveness of industrial operations [6]. This also explores the multifaceted need for C&I in the industry environment, emphasizing its pivotal role in maintaining operational excellence as well shown in Fig. 6.2 [7].

One of the primary reasons for the widespread adoption of C&I in industries is the substantial improvement in operational efficiency it offers. Control

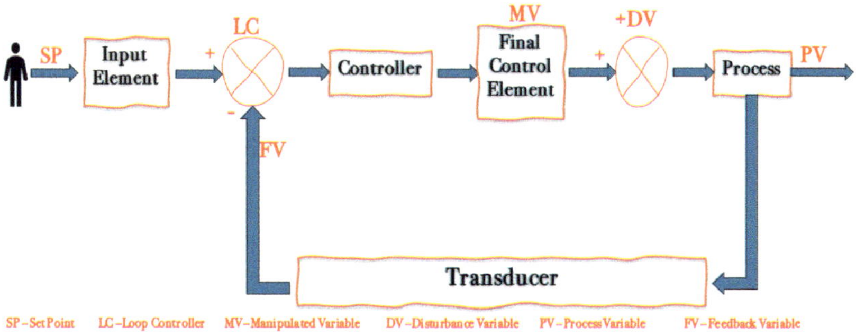


**FIGURE 6.2** Schematics of control and instrumentation system for switchgear operation.

systems enable automated regulation of various parameters such as temperature, pressure and flow, ensuring that manufacturing processes remain within predefined thresholds. This automation not only reduces the margin of error but also enhances the consistency and accuracy of production processes. In a competitive industrial landscape, where efficiency is synonymous with profitability, the implementation of C&I systems becomes a strategic imperative.

Instrumentation plays a crucial role in providing real-time monitoring and data acquisition capabilities, allowing industrial operators to make informed decisions promptly. Sensors and instruments are deployed throughout the manufacturing or processing plant to capture data on variables such as temperature, pressure, level and other critical parameters. This real-time data is then processed by control systems, enabling operators to monitor the health and performance of machinery, identify deviations from optimal conditions, and take corrective actions as needed [8]. The ability to gather and analyze data in real-time empowers industries to achieve higher levels of precision, predict potential issues and optimize their processes for maximum efficiency.

Safety is a paramount concern in any industrial setting, and C&I systems play a pivotal role in mitigating risks associated with complex processes. Automated control systems can swiftly respond to abnormal conditions, triggering alarms or initiating emergency shutdown procedures to prevent accidents or equipment failures. For instance, in a chemical processing plant, C&I systems can monitor and control factors such as temperature and pressure to prevent hazardous situations. The integration of safety interlocks and alarms ensures that operators are promptly alerted to potential issues, allowing for timely intervention, and reducing the likelihood of accidents.

Industries are constantly evolving to meet changing market demands and technological advancements. C&I systems offer a high degree of adaptability and flexibility, allowing industries to modify and optimize their processes without extensive manual intervention. Programmable logic controllers (PLCs) and distributed control systems (DCSs) enable the reprogramming of control strategies, making it easier for industries to adapt to new production requirements, implement process improvements or incorporate new technologies. This adaptability is crucial for industries striving to remain competitive and responsive to market dynamics.

In conclusion, the need for C&I in the industry environment is undeniable. The integration of sophisticated control systems and precise instrumentation is essential for enhancing operational efficiency, ensuring safety and maintaining competitiveness. As industries continue to face increasing complexity and global competition, the strategic implementation of C&I systems becomes not only a technological necessity but also a key driver of sustainable growth and success. The continuous advancement of C&I technologies promises to further revolutionize industrial processes, setting the stage for a future where smart and interconnected systems redefine the possibilities within the industrial landscape.

## 3. Speciality of automation in sub-station environment

Automation in the substation environment represents a specialized and critical facet of modern electrical infrastructure management. Substations serve as pivotal nodes in the power distribution network, and the implementation of automation technologies brings forth a range of advantages [9]. Automated systems in substations enable remote monitoring, control and operation, reducing the need for manual interventions in potentially hazardous environments. Smart sensors and sophisticated control algorithms ensure real-time data acquisition, facilitating the rapid identification of faults and abnormalities. Automated substation systems enhance grid reliability by enabling quick responses to changing demand patterns and disturbances, optimizing voltage levels and facilitating the seamless integration of renewable energy sources. Furthermore, automation contributes to preventive maintenance strategies, enhancing the overall efficiency and longevity of substation equipment. The speciality of automation in the substation environment lies in its ability to bring intelligence, resilience and efficiency to the core of power distribution networks, laying the foundation for a reliable and adaptive electrical infrastructure.

The need for standards such as IEC 61850/61851 and IEEE C37.1 stems from the increasingly complex and interconnected nature of modern electrical systems, particularly in the realm of power distribution and substation automation. Let us delve into each standard:

### 3.1 IEC 61850/61851

**IEC 61850:** This standard, developed by the International Electrotechnical Commission (IEC), addresses communication protocols and data models for substation automation systems. It defines a framework for the integration of various intelligent electronic devices (IEDs) within substations, enabling interoperability and standardized communication between devices from different manufacturers. IEC 61850 promotes the use of Ethernet-based communication networks, offering benefits such as higher bandwidth, flexibility and scalability compared with traditional serial-based communication protocols. The standard also facilitates the exchange of real-time data for functions such as protection, control, monitoring and maintenance, ultimately enhancing the efficiency, reliability and interoperability of substation automation systems.

**IEC 61851:** This standard focuses on electric vehicle (EV) charging infrastructure, specifying communication protocols and data formats for interactions between EVs and charging equipment. It ensures compatibility and interoperability among EVs and charging stations from different manufacturers, enabling seamless integration into existing electrical grids. IEC 61851 addresses aspects such as communication protocols, connector types, charging modes and safety requirements, thereby facilitating the widespread adoption and deployment of EV charging infrastructure worldwide [10].

### 3.2 IEEE C37.1

IEEE C37.1, developed by the Institute of Electrical and Electronics Engineers (IEEE), is a standard for the protection and control of power systems and equipment. It provides guidelines and requirements for the design, testing, installation, operation and maintenance of protective relays and associated devices used in electrical substations. IEEE C37.1 covers aspects such as relay functions, performance criteria, testing procedures, communication interfaces and interoperability standards. By establishing common practices and specifications for protective relaying equipment, IEEE C37.1 ensures the reliability, safety and effectiveness of protection schemes in power systems, thereby safeguarding equipment and personnel from electrical faults and disturbances.

In summary, the need for standards such as IEC 61850/61851 and IEEE C37.1 arises from the imperative to standardize communication protocols, data formats and equipment specifications in electrical systems and substations. These standards promote interoperability, reliability, efficiency and safety in the design, operation and maintenance of modern power distribution and substation automation systems, ultimately contributing to the stability and resilience of electrical grids and infrastructure.

### 3.3 Applications of the standards

IEC 61850 and IEC 61851 standards find application in distinct domains, addressing communication protocols and data models for substation automation and EV charging infrastructure, respectively [11]. Let us explore the specific applications of each standard:

1. **IEC 61850 – Substation Automation**:
   - **Interoperability in Substation Automation**: IEC 61850 ensures interoperability among various intelligent electronic devices (IEDs) within a substation. This includes protection relays, metres, switches and other devices from different manufacturers. By standardizing communication protocols, data models and system architectures, IEC 61850 allows seamless integration and communication between devices, fostering a more flexible and interoperable substation automation environment.
   - **Ethernet-based communication networks**: The standard promotes the use of Ethernet-based communication networks in substations [12]. This shift from traditional serial-based communication protocols to Ethernet enhances communication speed, bandwidth and scalability. Substations can utilize Ethernet networks for real-time exchange of data related to protection, control and monitoring, contributing to improved operational efficiency and reliability.
   - **Enhanced monitoring and control**: IEC 61850 facilitates the exchange of real-time data between devices, enabling advanced monitoring and

control functions. Substation operators can remotely monitor equipment status, diagnose faults and implement control actions more efficiently. This leads to faster response times in addressing abnormalities, reducing downtime and enhancing the overall reliability of the power grid.
- **Standardized naming conventions and data models**: The standard defines standardized naming conventions and data models, ensuring consistency in the representation and interpretation of data across different devices. This commonality simplifies engineering processes, reduces configuration errors and enhances the overall maintainability of substation automation systems.

2. **IEC 61851 – EV charging infrastructure**:
   - **Interoperability in EV Charging**: IEC 61851 addresses the need for interoperability among different EVs and charging infrastructure. The standard defines communication protocols, connector types and charging modes to ensure that EVs can be charged at various charging stations worldwide, regardless of the manufacturer.
   - **Safety and performance standards**: IEC 61851 establishes safety requirements and performance standards for EV charging equipment. This includes considerations for electrical safety, communication safety and mechanical safety to safeguard both the EV and the charging infrastructure. It also outlines the communication interfaces between the EV and the charging station to facilitate efficient charging and monitoring.
   - **Common charging modes**: The standard defines different charging modes, such as mode 1, mode 2, mode 3 and mode 4, each specifying the charging process, connector types and safety features [13]. These standardized modes ensure that EVs can be charged safely and efficiently across diverse charging infrastructures.
   - **Facilitating global EV adoption**: IEC 61851 plays a crucial role in promoting the widespread adoption of EVs by providing a standardized framework for charging infrastructure. This enables EV manufacturers and charging station operators to develop products that adhere to a common set of standards, fostering a more extensive and interoperable EV charging network globally.

In summary, IEC 61850 and IEC 61851 standards contribute significantly to the standardization, interoperability and safety of substation automation and EV charging infrastructure, respectively, ensuring the seamless integration and operation of devices and systems in these domains.

IEEE C37.1, a standard developed by the Institute of Electrical and Electronics Engineers (IEEE), focuses on the protection and control of power systems and equipment [14]. The application of IEEE C37.1 is primarily in the domain of protective relaying equipment and associated devices within electrical substations. The following are key applications of the IEEE C37.1 standard:

1. **Design and development of protective relays**:
   - IEEE C37.1 provides guidelines and requirements for the design and development of protective relays [15]. These relays play a critical role in power systems by detecting abnormal conditions, such as overcurrent, faults and other disturbances, and initiating protective actions to isolate the affected equipment and prevent further damage.
2. **Performance criteria for protective relays**:
   - The standard outlines performance criteria that protective relays must meet to ensure their effectiveness in detecting and responding to system faults. This includes parameters such as accuracy, speed of operation, sensitivity and selectivity. Compliance with these criteria is essential to guarantee the reliability and stability of the power system.
3. **Testing procedures and acceptance criteria**:
   - IEEE C37.1 provides detailed testing procedures for protective relays and associated devices. It includes acceptance criteria that help verify whether the protective relays meet the specified performance standards. This ensures that the protective relaying equipment is tested thoroughly before being deployed in actual power system applications.
4. **Installation and operation guidelines**:
   - The standard offers guidelines for the proper installation and operation of protective relaying equipment. This includes considerations for wiring, settings, coordination with other protective devices and environmental conditions. Adhering to these guidelines is crucial to achieving optimal performance and reliability of the protective relays in the field.
5. **Maintenance and inspection practices**:
   - IEEE C37.1 provides recommendations for maintenance and inspection practices to ensure the continued reliability of protective relaying equipment throughout its lifecycle. Regular maintenance and inspections are essential for identifying and addressing issues that may arise over time, preventing potential failures and ensuring the longevity of the protective relays.
6. **Communication interfaces and interoperability**:
   - While IEEE C37.1 primarily focuses on protective relaying equipment, it may also touch upon communication interfaces between relays and other devices in the substation. Interoperability considerations ensure that protective relays can communicate effectively with other elements of the substation automation system, contributing to a more integrated and efficient power system.

In summary, IEEE C37.1 plays a crucial role in guiding the design, testing, installation, operation and maintenance of protective relaying equipment within electrical substations. By providing a standardized framework and performance criteria, this standard contributes to the overall reliability, safety and effectiveness of protective relays in safeguarding power systems against faults and disturbances.

## 4. Data exchange between field smart devices and control centre

Data exchange between field smart devices and a control centre is a critical aspect of modern industrial automation and substation monitoring.

Various sensors play a crucial role in collecting and transmitting real-time data, ensuring the safe and efficient operation of power systems. Here, we will discuss the functionalities of specific sensors, the importance of avoiding humidity, real-time messaging from the human−machine interface (HMI) to activate protective relays and the standards/protocols involved in data communication.

1. **Gas density sensor**:
   - **Functionality**: Gas density sensors are often used in gas-insulated switchgear (GIS) to monitor the density of insulating gases such as sulphur hexafluoride (SF6). Changes in gas density can indicate potential faults or abnormalities. These sensors provide continuous monitoring, allowing for early detection and timely intervention to prevent equipment failures.
   - **Importance**: Early detection of gas leaks is crucial to maintaining the integrity of the insulation and preventing environmental and safety hazards. Gas density sensors contribute to the overall reliability of GIS by providing real-time information on gas conditions.
2. **Avoidance of humidity**:
   - **Functionality**: Humidity sensors are employed to monitor and control the moisture levels in critical equipment, especially in environments where insulation materials are sensitive to moisture. High humidity can lead to reduced insulation performance and increased risk of electrical breakdown.
   - **Importance**: Avoiding humidity is essential for maintaining the dielectric properties of insulation materials. Humidity sensors help prevent moisture-related issues, ensuring the reliability and longevity of electrical equipment.
3. **Real-time messaging from HMI to activate protective relays**:
   - **Functionality**: Real-time messaging from the HMI to field devices involves the instant exchange of information between the control centre and smart devices in the field. This communication is crucial for controlling and monitoring equipment, activating protective relays and responding swiftly to changing conditions.
   - **Importance**: Real-time communication enables operators to make informed decisions promptly. For example, if a protective relay needs to be activated in response to a detected fault, real-time messaging ensures that the relay is tripped without delay, preventing further damage to the system.

4. **Standards/protocols involved in data communication**:
   - **IEC 61850**: This standard plays a significant role in defining communication protocols and data models for substation automation. It ensures interoperability between different devices, including sensors, relays and HMIs, by providing a common framework for communication and data exchange.
   - **DNP3 (distributed network protocol)**: Widely used in the utility industry, DNP3 is a protocol that facilitates communication between various devices, including sensors and control centres. It supports real-time data exchange and is designed for reliable operation in challenging environments.
   - **Modbus**: Modbus is a widely adopted protocol for industrial communication. It is often used to facilitate communication between sensors and control systems. Modbus supports both serial and Ethernet communication, making it versatile for different applications in the field of industrial automation.

In conclusion, the seamless exchange of data between field smart devices and control centres involves the utilization of specialized sensors, careful monitoring of environmental conditions and the implementation of standards and protocols such a IEC 61850, DNP3 and Modbus as shown in Fig. 6.3 [16]. These elements collectively contribute to the efficiency, reliability and safety of systems by enabling real-time monitoring, control and timely response to dynamic conditions in the field.

## 5. Future aspects of substation automation

The future of substation automation holds several exciting developments, incorporating advancements in standards, industrial Internet of things (IIoT) and innovative sensing technologies.

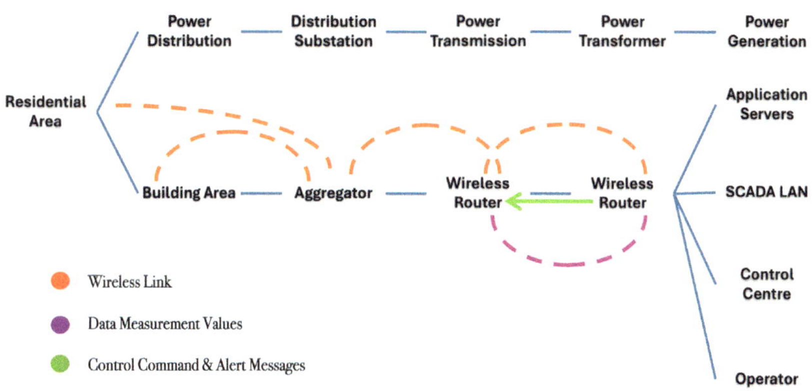


FIGURE 6.3 Communication link in a cyberphysical system.

Here are some future aspects and trends in substation automation:

1. **IEEE 1451.5 to IEEE 1451.7 standards**:
   - **Sensor interoperability**: The IEEE 1451 family of standards focuses on smart transducer interfaces [17]. IEEE 1451.5 specifies a standardized TEDS (Transducer Electronic Data Sheet) for sensors, facilitating plug-and-play interoperability. Future developments may involve the integration of IEEE 1451.5 and IEEE 1451.7, which extends these capabilities to wireless sensor networks, enabling seamless integration of various sensors in substation automation systems.
2. **Industrial IoT (IIoT)**:
   - **Integration of IIoT Platforms**: Substation automation is expected to leverage the capabilities of IIoT platforms for enhanced connectivity, data analytics and real-time decision-making. IIoT enables the integration of devices, sensors and systems into a unified ecosystem, providing more comprehensive insights into substation operations, predictive maintenance and overall grid health.
3. **Adaptation of sensing technology for new insulators**:
   - **N2SF6 and C3H7N insulators**: As the energy industry continues to explore alternative insulating gases for eco-friendly and sustainable solutions, sensing technologies will play a crucial role in monitoring and ensuring the reliability of new insulators such as nitrogen—sulphur hexafluoride (N2SF6) and propane (C3H7N) [18]. Advanced sensors will be employed to monitor gas composition, pressure and other parameters, ensuring the integrity of these new insulation materials.
4. **Machine learning and artificial intelligence (AI)**:
   - **Predictive analytics for maintenance**: The integration of machine learning and AI algorithms in substation automation will enable predictive analytics for maintenance. By analyzing historical and real-time data from sensors and devices, these technologies can predict potential equipment failures, optimize maintenance schedules and enhance the overall reliability of substation infrastructure.
5. **Cybersecurity enhancements**:
   - **Securing intelligent substation networks**: With the increasing connectivity and digitization in substation automation, there will be a heightened focus on cybersecurity. Future developments will involve implementing robust cybersecurity measures to protect intelligent substation networks from potential cyber threats, ensuring the integrity and confidentiality of critical data.
6. **Decentralized control architectures**:
   - **Edge computing in substations**: The adoption of edge computing in substation automation will lead to more decentralized control architectures. Edge devices, located closer to sensors and actuators, will process data locally, reducing latency and enhancing the responsiveness of control systems. This shift can improve system resilience and efficiency.

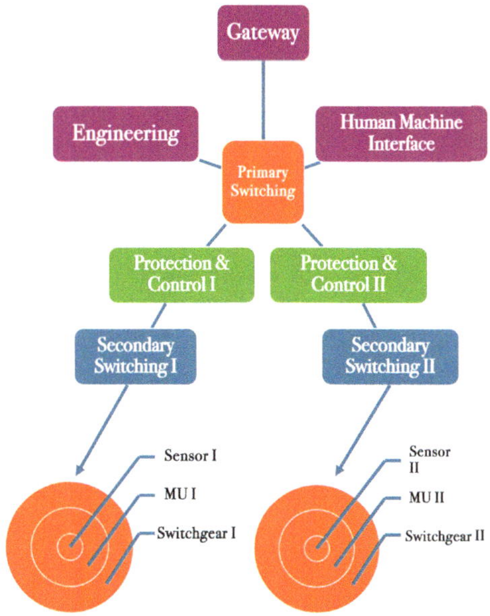


**FIGURE 6.4** Control aspects of substation automation.

7. **Advanced grid monitoring and control**:
   - **Wide-area monitoring and control (WAMC)**: Future substation automation systems will likely include advanced wide-area monitoring and control functionalities [19]. This involves the integration of synchrophasor technology and other wide-area measurement systems to enhance situational awareness, grid stability and the ability to respond to dynamic grid conditions in real time.

As substation automation continues to evolve, the convergence of standards, IoT technologies, advanced sensing capabilities and intelligent data analytics will shape a more resilient, efficient and sustainable power grid. These future developments aim to address the growing complexities of modern energy systems while ensuring the reliability and security of the critical infrastructure. There is also a dire need to strengthen the control aspects of the substation automation as shown in Fig. 6.4 [20].

## 6. Some brief industrial case studies on control and instrumentation aspects

### 6.1 Case 1: Efficient monitoring and control of switchgear using advanced instrumentation

In a modern manufacturing facility, optimizing energy consumption and ensuring the safety of electrical equipment is a top priority. The facility installed advanced instrumentation in its switchgear systems, including sensors for voltage, current and temperature monitoring. These instruments provided real-time data, allowing the control system to proactively identify any deviations from the normal operating conditions.

By continuously monitoring these parameters, the control system was able to detect anomalies and potential faults at an early stage. This early detection enabled the facility to schedule maintenance during planned downtime, preventing costly unplanned outages. Moreover, the system's ability to monitor temperature trends helped prevent overheating and ensure the safety of the switchgear equipment.

## 6.2 Case 2: Automation and remote monitoring in switchgear control

A power distribution company faced challenges in managing its switchgear systems across multiple substations. By implementing remote monitoring and control technologies, the company was able to centralize its operations and monitor switchgear across various locations.

Advanced communication protocols such as Modbus and IEC 61850 were used to integrate the switchgear with the central control system. This allowed the company to remotely operate breakers, monitor fault status and assess the condition of switchgear components. The automation reduced the need for on-site visits, saving time and resources.

Additionally, the company could quickly isolate faulty sections during an outage, minimizing downtime and improving the overall reliability of power distribution.

## 6.3 Case 3: Integration of smart sensors for predictive maintenance

An industrial plant sought to improve the reliability of its switchgear and minimize unexpected failures. The plant integrated smart sensors into its switchgear panels to monitor critical parameters such as insulation resistance, contact wear and partial discharge.

These smart sensors fed data into a predictive maintenance system that analyzed the condition of the switchgear components. The system generated alerts when it detected signs of wear or potential failure, allowing maintenance teams to take corrective actions before a fault occurred.

This approach not only extended the life of the switchgear equipment but also enhanced the plant's operational efficiency and safety. By leveraging smart instrumentation, the plant achieved a more proactive maintenance strategy and reduced the risk of costly outages.

## 7. Let us address these real-life issues arising on C&I aspects of switchgear

The technical know-how obtained from this chapter will help in enlightening some real-life industrial issues along with exploring their possible efficient solutions.

1. How do modern control systems integrate with legacy switchgear equipment, and what challenges arise during the integration process?
2. What role does real-time monitoring play in the safe and efficient operation of switchgear systems, and how can this data be leveraged for predictive maintenance?
3. What are the key differences between conventional and smart switchgear instrumentation, and how do these differences impact system performance and safety?
4. How can advanced control and instrumentation technologies in switchgear improve energy efficiency and reduce operational costs?
5. What are the latest developments in communication protocols for switchgear control, and how do they enhance interoperability and data exchange?
6. In what ways can the use of IoT-enabled sensors in switchgear systems contribute to enhanced remote monitoring and management capabilities?
7. What measures can be taken to ensure the cybersecurity of control systems in switchgear, especially when integrating with external networks?
8. How does the integration of advanced protection relays impact the reliability and safety of switchgear operations?
9. What strategies can be employed to minimize arc flash incidents in switchgear systems using advanced control and instrumentation techniques?
10. How do environmental factors such as temperature, humidity and pollution affect switchgear performance, and how can advanced instrumentation help mitigate these impacts?

## References

[1] G.R. Biswal, Role of Reliable Cooling System in Combined Cycle Power Plant towards Green Energy, Electrical India, Chary Publications, January 2013, pp. 94–100.
[2] A. Meier, Electric Power Systems: A Conceptual Introduction, Wiley Interscience, 2006.
[3] B. Ravindranath, M. Chander, Power System Protection and Switchgear, New Age International (P) Ltd, 2005.
[4] R. Schuerger, R. Fleig, O. Escobedo, AC vs. DC over-potential testing of AC equipment, in: IEEE/IAS 52nd Industrial and Commercial Power Systems Technical Conference, IEEE, Detroit, 2016.

[5] A. Silos, A. Senis, R.M. Pozuelo, A. Zaballos, Using IEC 61850 GOOSE service for adaptive ANSI 67/67N protection in ring main systems with distributed energy resources, Energies 1685 (2017).

[6] R. Smeaton, Switchgear and Control Handbook, McGraw-Hill, 1987.

[7] H. Mamur, et al., Modelling and Application of a Computer Controlled Liquid Level Tank System 2017. 7, CS & IT Conference Proceedings, 2017.

[8] A. Srivastava, K.G. Ravikumar, G. Zweigle, Wide-area monitoring and control using the real time digital simulator and a synchrophasor vector processor, Eur. Trans. Electr. Power 21 (4) (2010) 1521–1530.

[9] A. Jarosz, PID Control Approach for Optimizing Voltage Regulation in Smart Grids Using Energy Storage Reactive Power 2024, 2024.

[10] J. Kempf, L. Bai, M. Pipattanasomporn, M. Kuzlu, U. Cali, Digitalization of Power Markets and Systems Using Energy Informatics, Springer International Publishing, 2021.

[11] V. Vlad, C. Popa, R. Pentiuc, C. Buzduga, Iasi Control Architecture for Power Distribution Systems Based on IEC 61850, IEC 61499 and Holonic Concepts 2014, 2014.

[12] G. Jyothi, B. Kumar, C. Ananda, K. Nisha, Design and implementation of livetap for deterministic ethernet bus using FPGA, in: IEEE International Conference on Recent Trends in Electronics, Information and Communication Technology (RTEICT), Bangalore, 2016.

[13] IEC 62271—High-voltage, Switchgear and Control Gear Standards by the International Electrotechnical Commission.

[14] IEEE Std C37.20.1—IEEE Standard for Metal-Enclosed Low-Voltage Power Circuit Breaker Switchgear.

[15] T. Zhao, L. Sevov, College Station IEEE Practical Considerations of Applying IEC61850 GOOSE Based Zone Selective Interlocking Scheme in Industrial Applications 2013, 2013.

[16] B.A.V. Jha, Smart grid cyber-physical systems: communication technologies, standards and challenges, Wireless Network (2022) 2595–2613.

[17] E. Song, K. Lee, Understanding IEEE 1451-Networked smart transducer interface standard—what is a smart transducer? IEEE Instrum. Meas. Mag. 11 (2) (2008) 11–17.

[18] A. Hauashdh, S. Nagapan, J. Jailani, Y. Gamil, An integrated framework for sustainable and efficient building maintenance operations aligning with climate change, SDGs, and emerging technology, Results in Engineering 21 (101822) (2024).

[19] J. Cow, Wide-Area monitoring and control using the real time digital simulator and a synchrophasor vector processor, Eur. Trans. Electr. Power (2011).

[20] G.T. Saeid, G.M. Behrooz, IEC61850 implementation in localization of power substation automation system, in: 2014 IEEE International Symposium on Robotics and Manufacturing Automation (ROMA), KL, Malaysia, December 2014, pp. 92–98.

Chapter 7

# Testing and validation of smart and intelligent sensors

**Tarikul Islam[1] and Shailesh Kumar[2]**
[1]*Jamia Millia Islamia (A Central University), Department of Electrical Engineering, Delhi, New Delhi, India;* [2]*Arka Jain University, Department of Electrical and Electronics Engineering, Mohanpur, Jharkhand, India*

## 1. Introduction

Switchgear is the main component of an electric power system, It is composed of circuit breakers, fuse and electrical disconnect switches which are used to isolate, protect and control the electrical equipment. Switchgear is used to both to de-energize equipment and clear downstream faults. The electric power supply reliability is directly linked to such types of equipment. The predictive maintenance of switchgear is essential to maintain the continuous supply in the power system. Numerous physical parameters such as temperature, humidity and pressure need to be monitored and controlled inside the switchgear. The selection of a suitable sensor for the reliable measurement of the physical parameters of switchgear is the major challenge for its predictive maintenance. Further, the sensors must withstand the extreme environmental conditions of switchgear. The lack of measurement data for predictive maintenance of the switchgear is another challenge [1]. The level of temperature humidity and pressure around the switchgear should be continuously monitored and controlled for its efficient operations. The continuous measurement of temperature and humidity inside the switchgear vessel is very rare, or even nonexistent to maintain its long lifetime. Switching operations a few times a year are performed to monitor the breaker drive for maintenance purposes. To develop the AI/ML algorithms for smart monitoring and controlling the switchgear, measurement data are rarely available. In breaker drive monitoring, the situation is further exacerbated because of the fast response time demand in tens of milliseconds which makes reliable prediction very challenging. Nowadays, sulphur hexafluoride ($SF_6$)-gas-insulated switchgears (GISs) are used in transmission substations due to their frequent maintenance requirements, soaring land cost and compact design in urban areas [2,3]. The

**Switchgear Design, Operation, and Maintenance Using Industry Standards**
**https://doi.org/10.1016/B978-0-443-24718-7.00007-X**

excess moisture content in the SF6-GIS environment creates a critical scenario for the live electrical equipment and conductors [4]. Moisture originates in the surrounding area due to sudden temperature changes. It occurs due to arching taking place in response to switching actions which activates required protective mechanisms [5–7]. The selection of the sensor for the measurement of a certain level of humidity depends upon its electrical characteristics. For high precision and accuracy, the cost of a moisture sensor detecting the ppm level moisture is in the range of thousands of dollars [8].

Studies have shown that the life of the transformer depends upon the mechanical, chemical and electrical strength of its insulator. Over the period, the life of the insulator degrades. The most common cause of this degradation is the moisture originating inside the transformer and ingress from the environment atmosphere [9]. During the normal operation of the transformer, the temperature of oil changes causing expansion and contraction of oil inside the transformer. Due to the temperature gradient of the transformer, the atmospheric air is sucked from the transformer through the breather as the temperature of insulating oil decreases, and air available at the top surface of the oil comes to the atmosphere when the temperature decreases. This process is known as the breathing of the transformer [10,11]. Ingresses of atmospheric moisture inside the transformer are the major part of moisture contamination. Moisture contaminates oil in the transformer, and a myriad of negative effects start occurring. For example, moisture leads to chemical, electrical and mechanical strength determination of the oil, which in turn leads to particulate contamination. The viscosity of the oil also changes due to the moisture and causes sludge formation [12]. To protect transformers against excess atmospheric moisture contamination, there is a need for a holistic contamination control approach. The desiccant used in the breather is hygroscopic. It attracts and holds the moisture molecules and maintains dryness in its vicinity. Transformer breather commonly uses different types of silica gel as a desiccant material [13]. For this reason, a desiccant breather is used to filter the atmospheric air moisture and particles during the breathing of the transformer. It is connected to the conservator of the transformer through the breathing pipe. The moisture-absorbing capacity of the silica gel breather deteriorates over time [14]. Nowadays, a lot of instrumentation techniques are used to know the quality of the insulator system of the transformer so that preventive action can be implemented before the occurrence of the fault. It can avoid the unwanted occurrence of the transformer fault and improve its operational life [15]. The literature shows that the application of sensors for accurate monitoring of the physical parameters such as humidity temperature, pressure gas density level of the gases, etc. became very important, particularly in switchgear and transformer health monitoring.

A normal sensor can detect physical parameters such as flow, temperature, pressure, etc. The output of the sensor is related to the detected physical parameter. For example, switching indication is the output of the sensor. The input and output of the sensor are analogues to each other which indicates the continuous monitoring of the input parameter. The output of the moisture sensor is measured as the change of impedance/capacitance/resistance or volt with the variation of the level of relative humidity (RH) around it based on its principle of operation. A normal sensor itself is not enough to be used in the automation system. A transducer is used to convert the change in capacitance or resistance value into a useable signal level in the range of 4–20 mA value. A computing facility is required to utilize such measured value in industrial automation for performing some calculations and taking any corrective action. A smart sensor has an in-built computing facility for the analysis of the measured physical parameters to improve the efficiency of the automation process. The output of the smart sensor is more accurate. The built-in computing resources inside the sensor filter the noise signal. The measured signal is converted into a digital format without any use of a transducer. Due to the built-in communication facility inside the smart sensor, the measured data can be transferred to the Internet or other network for further processing. As shown in Fig. 7.1, it consists of five basic elements:

**(i)** Sensor
**(ii)** Power supply
**(iii)** Microcontroller
**(iv)** Memory
**(v)** Communication module

The sensor is used to detect the physical parameters. Microcontroller is the computing resources. It enables the sensor to do calculations on measured physical quantities and take action based on the analysis. Memory is used for storing the measured physical parameters by the sensor and calculated data, and it also stores the logic of the incorporated software for controlling the sensor data. Through the communication module, the sensor data can be transferred to another device through the internet.

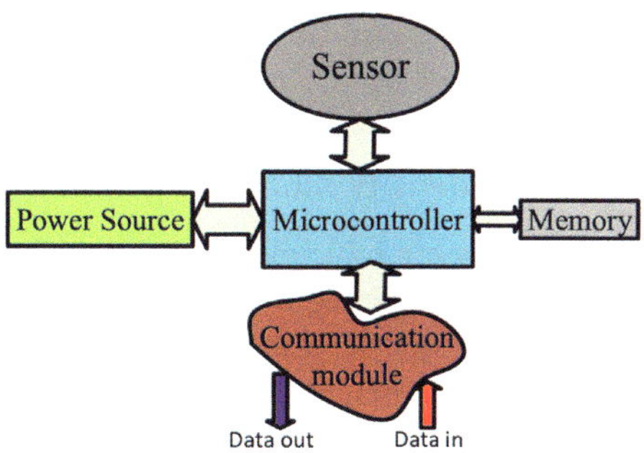


**FIGURE 7.1** Components of smart sensor.

## 2. Sensors used in monitoring the physical parameters in switchgear

### 2.1 Temperature sensors based on fibre optic

A direct contact type fibre optic sensor is preferred to install for the temperature measurement. This contact is made possible through the ring lug which is attached with fibre at its end point. A high-quality plastic of glass is used as a fibre. Fibre is isolated from electrical noise or events. It runs back to the temperature transmitter which is located in the low-voltage cabinet to a monitoring system. The advantages of such types of sensors in continuous monitoring of switchgear devices are as follows [16]:

- The fibre optic-based sensor is free from partial discharge, shorts and arc flashes.
- Measurement is highly reliable due to the wired connection at the measurement points.
- It is free from radio frequency interference and electromagnetic.
- For the real-time monitoring of the temperature, it can be installed underneath of the insulating boots.
- The response of the sensors is highly reproducible.

### 2.2 Moisture measurement in $SF_6$-GIS using capacitive humidity sensor

The designed sensor is a parallel plate capacitive type. Fluorine-doped tin oxide (FTO) glass acts as the lower plate and substrate, where the silver plate is the second and upper electrode of the designed sensor. A porous dielectric $Al_2O_3$ prepared through the sol-gel method is sandwiched between the parallel plate and acting as a moisture-sensing layer for $SF_6$ gas. The experiment was performed to measure the moisture in $SF_6$ gas. Linear characteristics and sensitivity of the sensor were found to be (±2.05%) and (0.238 pF/ppm), respectively, in the moisture range of 250–750 ppm. The hysteresis and response of the designed sensor were (±0.32%) and (±0.84%), respectively, with reputability of (±0.013%) [17].

### 2.3 Ultrasonic sensor for density monitoring of $SF_6$ gas

The sensor works on the ultrasonic principle. The time of flight and velocity of wave transverse in a chamber containing $SF_6$ gas are different than those of an empty chamber. $SF_6$ gas is acting as a medium for the ultrasonic wave. It relates to the simultaneous measurement of pressure and temperature inside the chamber using a manometer and temperature sensor relevantly with other measurement techniques [18].

## 3. Sensors used to monitor the electrical parameters in switchgear of electrical power systems

Voltage and current are the main electrical parameters in switchgear, which need to be monitored through the sensor.

### 3.1 Current sensor

The current flowing through the conductor is measured through the sensor. The magnitude of the current is measured in terms of current, voltage, or digital signal, which can be further utilized in various applications such as control, monitoring, or protection. Nowadays, two different types of current sensors (current transformers and Hall effect sensors) are used for the measurement of current in switchgear operations.

#### *3.1.1 Current transformers*

The CT consists of three parts: (1) main winding, (2) secondary winding, and (3) core. It is used for measuring the high alternating current. Through this device, a large current available across the primary winding is transformed into a smaller current in the secondary winding. CT is a passive device. It does not need an additional circuit to perform its operation. It can be used to monitor very high currents while using very little power. The core of CT is made of ferrite material. It can get saturated by a large amount of DC component in the main current which is why the signal at the secondary side of the CT is distorted. Due to the magnetic behaviour in the core material, it causes hysteresis, which lowers its accuracy. That is why it needs demagnetization. It works on the principle of electromagnetic induction. The change in flux around the core is proportional to the current change. CT cannot be used for detecting the DC. The schematic diagram of the CT sensor is shown in Fig. 7.2.

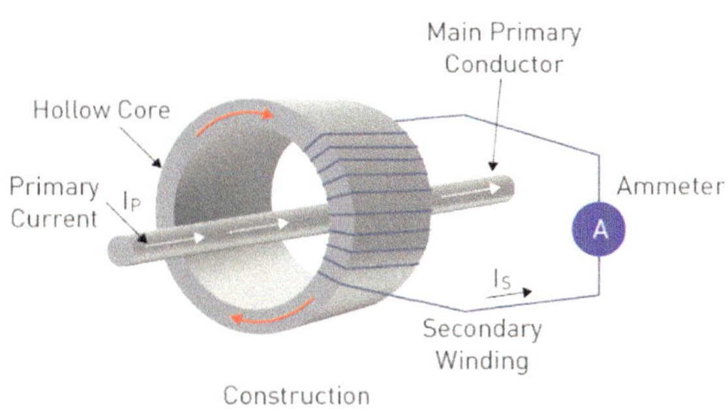


**FIGURE 7.2** Schematic diagram of CT sensor.

### 3.1.2 Hall effect sensors

These sensors work on the principle of the Hall effect. It states that a potential difference appears across the current caring conductor in the presence of a transverse magnetic field and cross section of the conductor. The schematic image of the sensor is shown in Fig. 7.3. The direction of the electromotive force (emf) is found to be perpendicular to the magnetic field and current flowing through the conductor. The amplitude of the phase difference between the current and magnetic field decides the result of vector voltage, and it is proportional to the Hall constant. The magnetic field around the conductor is developed due to the current flowing through the conductor and also provides a voltage that may be analyzed. A signal conditioning circuit is required to make the output useable in most applications. Generally, it requires temperature correction and an amplifier stage. A differential amplifier with these properties may simply be combined with the Hall element using standard BJT-based technology.

## 3.2 Voltage sensors

A voltage sensor is a passive device and is used for measuring the voltage in any electrical circuit. The ways of measuring voltage are different. The level of current is very low for measuring the high voltage. Such electrical devices are used for numerous applications such as power systems and industrial controls.

### 3.2.1 How does a voltage sensor work?

*Electromagnetic*: These types of sensors utilize electromagnetic fields to sense the variation in voltage level. A magnetic field is generated when a sensor is exposed to current. Such currents are due to the current flowing through the circuit board or wires and near the conductor. The sensor is found to be highly sensitive to detect the variation of current due to such types of elements. Such types of sensors are built with the microcontroller to detect the change of electromagnetic field around it due to analog-to-digital converters.

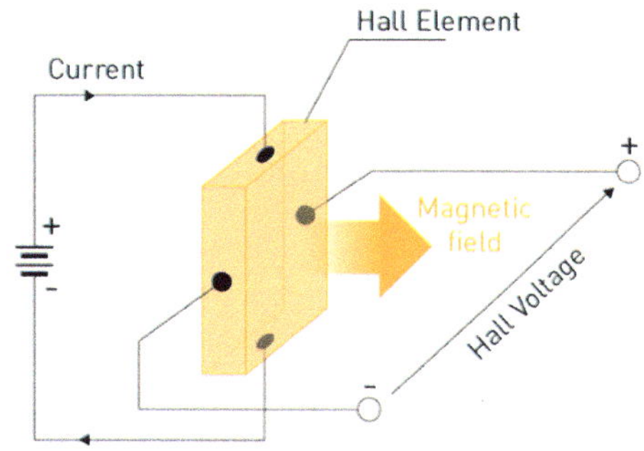


**FIGURE 7.3** Schematic diagram of Hall effect sensor.

### *3.2.2 Different types of voltage sensors*

Voltage sensors are of two different types: (1) measure voltage and (2) measure current. The voltage-sensitive resistor (VSR) is used to measure the voltage, and the operational amplifier is used to measure the current. A voltage-sensitive resistor is used to sense the small voltage in electronic circuits, such as for measuring the life of the battery and the output of the solar panel. Op-amps may be utilized for controlling LEDs and driving motors. It is also found inside the microcontrollers.

## 4. Desiccant breather for transformer applications

### 4.1 How desiccant breathers work

The design and construction of the breather are different for different types of transformers. The high-rated transformer (>500 MVA) uses a three-stage design of the breather to ensure the dry condition of the interior design of the transformer conservator. Fig. 7.4 shows the working of the transformer breather like this [14–16,19].

1. As the transformer breathes, the atmospheric air enters into the desiccant breather and passes through the particle filter. Through this process, the atmospheric air particles filter out two microns or greater.

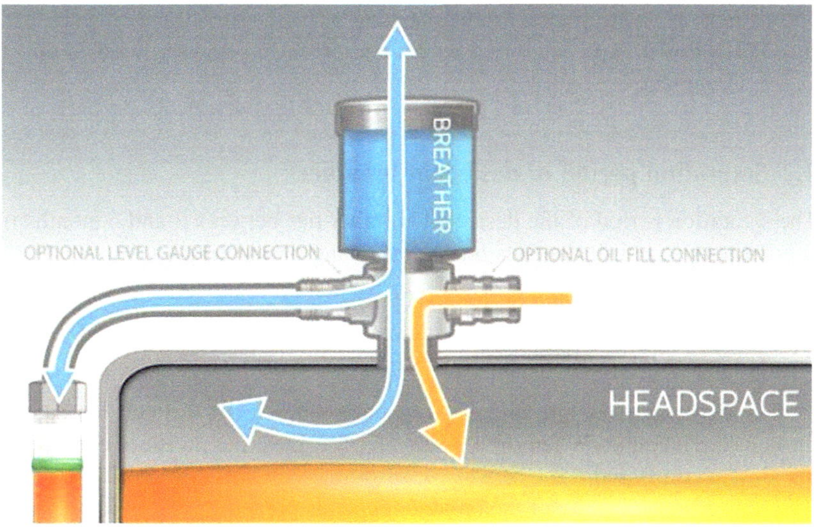


**FIGURE 7.4** Graphical presentation working model of breather with the conservator of distribution transformer.

2. The water molecule present in the atmospheric air is filtered through the silica gel.
3. Lastly, the atmospheric air passes through a third solid particle filter (micron-level) at the inlet of the conservator.

As the transformer exhales, the air flows inside the breather in a reverse direction and is purged to the atmosphere. The silica gel used in the breather is saturated in a short time. The colour of the silica gel gradually changes with its moisture-absorbing capacity. In the dry condition of the silica-gel, its colour is deep blue and changes to purples when gets saturated. It indicates its hydration or replacement time. Some additional components (carbon-filled foam, check valve and reusable tops) are integrated with the breather to expand its life. Carbon foam is a filter that is used to absorb the oil vapour at the bottom of the breather and protect the silica gel from the oil splashing up. Check valves and tops provide a closed loop to the breather if the airflow is needed when the pressure gradient between the atmosphere and inside the transformer exceeds the threshold level. Various breather models add the following components in addition to the check valve and tops.

(i) **High-capacity air filter**: It is used to protect the breather desiccant from dirty and dusty environmental conditions.
(ii) **Vacuum gauges**: It is the visual indicator of the replacement of the breather when the first particle filter is clogged with dust and debris.
(iii) **Isolation check valve**: It keeps the desiccant coming in contact with exhaust air and protects from splashing oil.
(iv) **Wash-down caps**: It is used where equipment is regularly washed dusty environment.

## 4.2 Saturation period of desiccant breather

The saturation period of the desiccant breather lies between 3 and 6 months if it is properly sized. It depends on the following parameters:

(i) Air intake volume and its frequency.
(ii) The capacity of a breather to accommodate moisture content in silica gel.
(iii) The water molecule available in atmospheric air.

The air intake volume and its frequency refer to how much air passes through the breather. When the transformer breathes, the moisture content of the air is retained in the gel, steadily deteriorating the life of the desiccant. The water-absorbing capacity of the breather depends upon the quality of the silica gel inside it. Atmospheric moisture not only deteriorates the life of desiccant, but also reaches the saturation level faster. When the saturation level of the desiccant is achieved, it does not absorb the more moisture content of the air.

## 4.3 Possible ways to improve the life span of transformer desiccant breather

(i) **Holding the breather at a proper height to the conservator**: Increase the height of the bottom of the breather to the top of the transformer conservator. A short-length pipe between the breather and the conservator should be added instead of directly attached to the conservator so that the breather can stay clear of moist air.
(ii) **Use of check valve and bladder**: The breather should assemble with the check valve and bladder so that it can be protected from the moist air coming from the conservator, which will improve the life span of the desiccant breather.
(iii) **Use of programmable heater in dehydrating breather**: The moisture saturation capacity of the desiccant (silica-gel) deteriorates over time. The rate of saturation depends upon the temperature gradient of transformer oil and the geographical environmental condition where the breather is installed. The moisture content of breather air depends upon the saturating condition of the desiccant. To avoid the manual inspection and frequent replacement of desiccant after achieving its saturation level, a breather should be installed with a programmable heather to dehydrate the desiccant at the pre-set threshold value of its saturation level. It will improve the operating efficiency of the transformer and the life span of the breather.

## 4.4 Size selection of desiccant breather for transformer

The size selection of the breather consists of the following factors:

(i) Rating of transformer
(ii) The level of air inside the conservator above the oil
(iii) The quality of the oil used in the transformer for the insulation purpose
(iv) Breathing rate of the transformer which is why the temperature of the transformer drops by 8°C in 24 h

Many of the desiccant breather models used in transformers involve multiple sizes in series. The selection of the breather size and its cycle time depending upon the important factors of the transformer are shown in Table 7.1.

## 4.5 Setting of desiccant breather to work for transformer application

Installation of a breather is a very simple process. The breather is mounted on the end of the connecting pipe coming from the conservator. However,

**TABLE 7.1** Parameters for the selection of transformer breather [13].

| Capacity of breather (kg) | Oil contents of transformer (L) of | Transformer rating (MVA) | Space for air volume in transformer(L) | Approximate daily breathing rate(L) of transformer | Volume of air prior to recharge (L) | Cycle time (Days) of breather |
|---|---|---|---|---|---|---|
| 0.60 | 1500 | <1.25 | 230 | 15 | 6000 | 400 |
| 1.20 | 3000 | 3 | 500 | 30 | 12,000 | 400 |
| 1.90 | 4750 | 6 | 1100 | 55 | 20,000 | 400 |
| 5.00 | 11,350 | 9 | 2400 | 140 | 45,450 | 320 |
| 10.00 | 22,700 | 30 | 4800 | 280 | 90,900 | 320 |
| 14.50 | 34,050 | 60 | 7250 | 420 | 136,500 | 320 |
| 17.50 | 45,450 | 100 | 9685 | 570 | 181,800 | 320 |
| 27.30 | 68,150 | 180 | 14,500 | 840 | 272,700 | 320 |

before use, visually check that no damage happened to any parts and the condition of the silica gel (if the gel is partially or fully saturated please proceed with the replacement or regeneration of the same). The setting of the breather to work for the transformer application is shown in image form in Fig. 7.5.

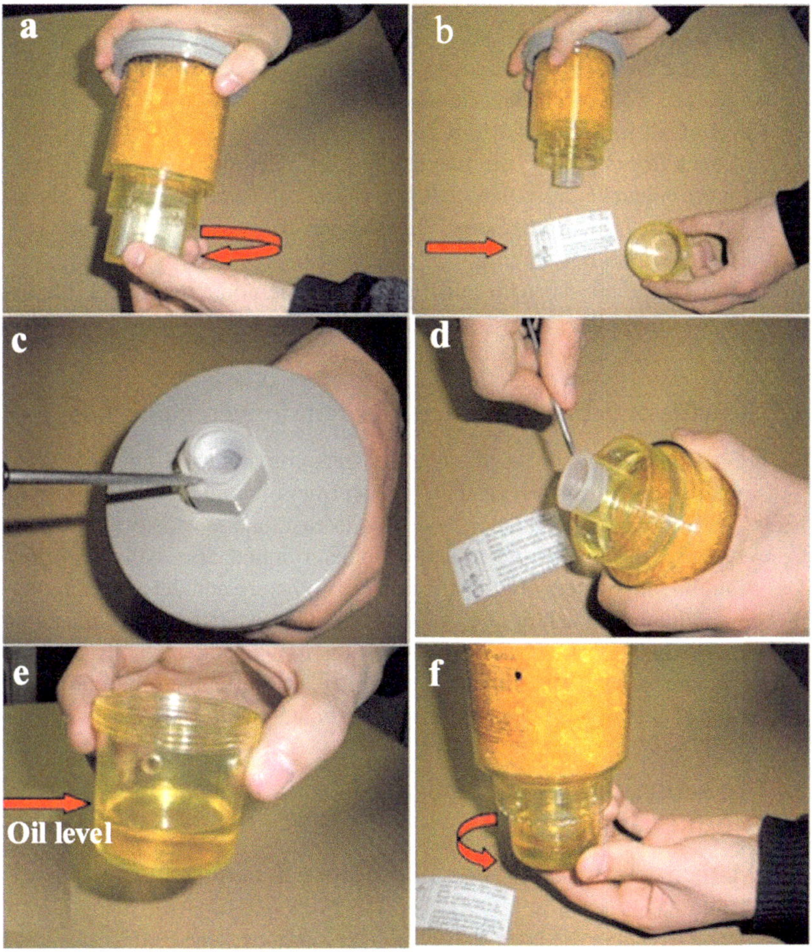


**FIGURE 7.5** (A) Unscrew the *bottom* cap, (B) take the instructions sheet out and read it carefully, (C, D) take out the plugs helping yourselves with a screwdriver, (E) fill the *bottom* cup with mineral oil up to the notch marked on cup, (F) screw in *bottom* cup – breather is now ready to be mounted with the conservator of the transformer. Setting of the breather to work for the transformer application.

### 4.6 Maintenance of desiccant breather for transformer application

To check the condition of the breather during maintenance of transformer is a good practice. First of all, the surface of the breather is cleaned, and then a few visual tests are performed:

- **Integrity of breather**
  The breather should not have visible external damages; gel container must be free of damages/cracks.
- **Saturation of silica gel**
  Check the colour of the silica gel: if a colour has changed from orange to white, you have to change the silica gel.
- **Level of oil inside oil bottom cup**
  Check the level of oil inside the cup. The oil level should reach approximately the notch on the bottom cap, and in case the level is found lower than this point, then the cup operates as follows:
  - Unscrew the bottom cup
  - Fill up the oil cup with mineral oil until the notch marked on the cap
  - Reassemble the oil cup on the breath

### 4.7 Benefits of desiccant breather for transformer application

As previously discussed, the transformer gets contaminated by atmospheric moisture, dirt and dust as it breathes. Transformer oil contaminated with water leads to additive oil oxidation, depletion, rust and corrosion over time. Consequently, the life of the transformer deteriorated drastically with time. Desiccant breather filters out the atmospheric moisture and debris and reduces the contamination level of the transformer. Consequently, it prolongs the lifespan of the transformer. Many condition monitoring methods of transformer breathers have saved companies money by shortening the frequency of downtime of the transformer and keeping it contamination-free.

## 5. The saturation level of silica-gel breather for transformer application

### 5.1 Mathematical formulation of moisture dynamics inside the breather

The transfer of moisture between air and gel inside the breather is a very sluggish process, and it takes several hours to reach a steady state condition. Transfer of moisture from the air to gel is a transient process, which consists of three physical processes, namely, (1) moisture transfer from air to silica-gel (*s-g*), (2) diffusion of moisture through the entire volume of *s-g*, and (3) humidity sensor response inside the breather. The air above the oil inside the conservator pushes up through the breather as a load of the transformer rises. Similarly, as the load of the transformer falls very sharply, environmental air flows into the conservator via the breather. This process is called the breathing of the transformer. A mathematical model can be used to analyze the moisture dynamics of the breathing process of a transformer. This derived model equation will predict the moisture content of silica-gel at a different RH of the breather air.

## 5.2 Modelling of moisture transfer from air to silica-gel inside the breather

The oil and air temperatures of the transformer are almost equal, but the temperatures of the breather air and transformer air are different. A mathematical relation between the breather air and transformer air can be given by (7.1),

$$T_1 = \frac{P_1 \times V_1 \times T_2}{P_2 \times V_2} \tag{7.1}$$

where $V_1$ is the breather volume in $m^3$, $V_2$ is the main tank volume in $m^3$, $T_1$ is the breather air temperature in Kelvin, $T_2$ is the transformer oil temperature in Kelvin, $P_1$ is the breather air pressure in mm Hg, and $P_2$ is the conservator air pressure of the transformer in mm Hg.

Further, considering moist air inside the breather permittivity *($\varepsilon_r$)*, the capacitance value ($C$) of the sensor can be given by a standard equation of capacitance [19].

$$C = \frac{\varepsilon_0 \times \varepsilon_r \times A}{d} \tag{7.2}$$

where $A$ indicates the cross-sectional area of the parallel plate, and $d$ indicates the gap between the parallel plates.

For the moist air inside the breather, the relative permittivity of the air is determined by Eq. (7.3) [19].

$$\varepsilon_r = 1 + \frac{211}{T}\left(P + \frac{48P_s}{T}\text{RH}\right) \times 10^{-6} \tag{7.3}$$

where $T$ indicates the breather air temperature in kelvin, $P$ indicates the breather air pressure, $P_s$ and RH denote the saturated vapour pressure and relative humidity at temperature $T$ of the breather air.

Therefore, the migration of the moisture between the air of the breather and the conservator at any temperature gradient depends only on the pressure of the breather. The *T-P (temperature—pressure)* relations between the conservator and breather air are given in (1). Silica-gel absorbs the moisture of the

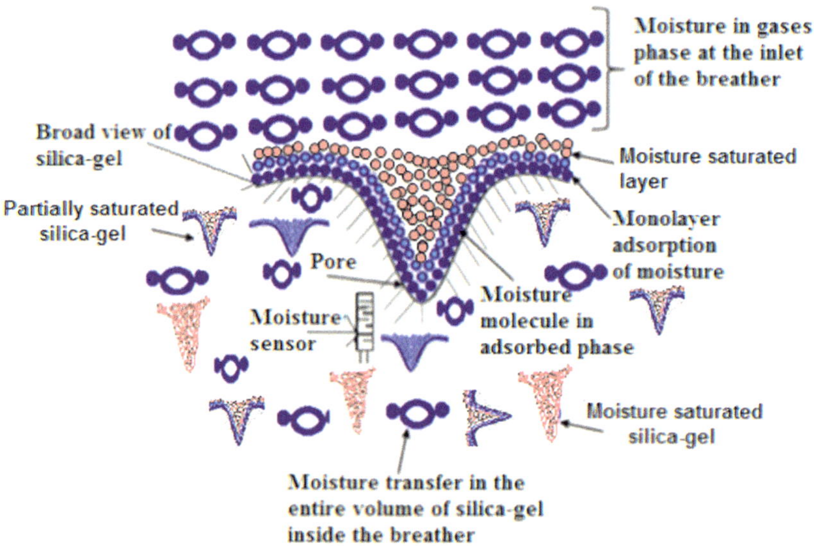


**FIGURE 7.6** Moisture transfer mechanisms from air to silica-gel inside the breather.

breathing air. The crystal structure of the silica-gel is different. It consists of large verities of pores in the range of macro- and micropores, and causes of non-uniform absorption of moisture by the gel. The micropore density of the silica-gel has an important role in the adsorption of moisture. The graphical presentation shows the transfer of moisture from the air to *s-g* in Fig. 7.6. It indicates the moisture distribution inside the pore of silica-gel for breather application. Initially, the moisture molecules are available in the breather air, and then gradually transferred to the gel, and gradually pores of the gel are filled up. The moisture equilibrium condition is achieved after a certain time elapsed in the adsorption process. The level of moisture inside the silica gel is indicated by the relative humidity measured by the RH sensor inside the breather. Here, moisture of the breather air is absorbed by silica-gel (*s*-g) at a particular temperature. A model was proposed by the Dubinin-Astakhov (D-A) for the migration of water in the crystals of silica-gel [41–43]. The equilibrium moisture concentration (EMC) in $cm^3/g$ is expressed by (7.4)

$$q = q_0 \exp\left[-\left(\frac{A}{E_0}\right)^{n_1}\right] + q_1 \exp\left[-\left(\frac{A}{E_1}\right)^{n_2}\right] \tag{7.4}$$

where $A$ indicates the adsorption capacity of gel at $T$ (temperature), $n_1$ and $n_2$ indicate structural heterogeneity of the micropores of the *s-g* (silica-gel), $E_0$ and $E_1$ represent moisture adsorption energy of the gel in cal/mole, $q_0$ and $q_1$ indicate limiting uptakes of adsorption space of the gel in $cm^3/g$.

The adsorption potential parameter $A$ may be given by

$$A = RT\ln\left(\frac{1}{\mathrm{RH}}\right) \tag{7.5}$$

where $R$ indicates the universal gas constant in J/Mol/K, $T$ and RH represent the temperature and relative humidity at equilibrium inside the breather. The equilibrium moisture concentration (EMC) is depicted as

$$\mathrm{EMC}\% = q \times 100 \tag{7.6}$$

An analogy exists between the heat and mass transfer of moisture [20]. Therefore, the process of heat and mass transfer of moisture may be correlated by Eq. (7.7).

$$\Gamma_{\mathrm{air}}\frac{\mathrm{dM}_{\mathrm{sgel}}}{\mathrm{dt}} + M_{\mathrm{sgel}} = M_{\mathrm{sgel-ss}} \tag{7.7}$$

where $M_{\mathrm{sgel}}$ indicates the mass of moisture at any time (t) and $M_{\mathrm{sgel-ss}}$ indicates the masses of moisture in silica-gel at the steady state condition.

$\Gamma_{\mathrm{air}}$ is the time constant of the air to maintain the moisture equilibrium with the silica gel at steady state conditions.

The mass of moisture content at a steady state $M_{\mathrm{sgel-ss}}$ changes as the temperature of the air inside the breather changes. The temperature-dependent value of $M_{\mathrm{sgel-ss}}$ for a certain period may be estimated by Eq. (7.7) and is modified as

$$\Gamma_{\mathrm{air}}\frac{dM_{\mathrm{sgel}}}{\mathrm{dt}} + M_{\mathrm{sgel}} = M_{\mathrm{sgel-ss}} + \overline{\Delta t}\frac{dM_{\mathrm{sgel-ss}}}{\mathrm{dt}} \tag{7.8}$$

where $\overline{\Delta t}$ indicates the small integration step.

## 5.3 Modelling of moisture transfer in the entire volume of silica-gel inside the breather

An RH (relative humidity) sensor is placed to monitor moisture inside the breather at a suitable place. During the breath-in process, the transformer sucks environmental air via a breather. During this process, atmospheric air moisture is uniformly distributed in the entire volume of *s-g* inside the breather. Now, relative humidity measured inside the breather is considered to be equal to the moisture mass inside the gel. This process may be formulated by Eq. (7.9).

$$M_{\mathrm{sens}}(t) = M_{\mathrm{sgel}}(t) \tag{7.9}$$

where $M_{\mathrm{sens}}(t)$ indicates the mass of moisture inside the *s-g*, which is measured by the RH sensor placed inside the breather.

## 6. Response of the RH sensor

The dynamic response is worked out to evaluate the concentration of moisture inside the *s-g* inside the breather. It can be analyzed using change in parametric values, that is, change in impedance/capacitance/resistance which is decided based on the working principle of the sensor [18]. Generally, a parallel plate capacitive sensor is used for measuring moisture in the form of a first-order differential equation [14]. The response of the sensor is given by

$$\Gamma_{\text{sens}}\frac{dM_b}{\text{dt}}+M_b=M_{\text{sens}} \tag{7.10}$$

where $M_b$ is the measured moisture inside the breather and $M_{\text{sens}}$ is the measured moisture inside the silica-gel.

After combining Eqs. (7.8)–(7.10), the final model equation is expressed as

$$\Gamma_{\text{air}}\Gamma_{\text{sens}}\frac{d^2M_b}{\text{dt}^2}+(\Gamma_{air}+\Gamma_{\text{sens}})\frac{dM_b}{\text{dt}}+M_b=M_{\text{sgel-ss}}+\overline{\Delta t}\frac{dM_{\text{sgel-ss}}}{dt} \tag{7.11}$$

Eq. (7.11) performs a dynamic response of the sensor using a mathematical model that indicates the concentration of moisture inside the breather at a steady-state condition.

## 7. Computational validation of the model

A developed model was validated in the frequency domain. A standard input signal indicating the real-time moisture variation inside the breather was used as input to validate the model. The input signal showing such type of variation of moisture inside the breather is shown in Fig. 7.7. Initially, for 15 days, the profile shows hyperbolic characteristics in a change in moisture concentration, and then it remains constant for the next 15 days (which is arbitrarily selected). The input indicates the situation where, for the initial 15 days, the moisture level increases sharply due to bad weather, then it remains constant for the next 15 days. At an equilibrium condition, the moisture concentration inside silica gel depends upon the RH inside the breather. In the selected input signal, X2 (t) = 0.98 for t > 15, which is the maximum moisture absorption capacity of the silica gel. It indicates the saturation level of gel, and beyond this humidity level, it is not useful for the further adsorption of moisture.

### 7.1 Development of transfer function of the moisture model inside the breather

Considering $M_{\text{sgel-ss}}$ is the input and $M_b$ is the output of the model Eq. (7.11). The response through Eq. (7.11) may be given by

$$Y(S)=\left[\frac{\overline{\Delta t}S+1}{P1'S^2+P2'S+1}\right]X(S)$$

where $P1'=\tau_{\text{air}}\tau_{\text{sens}}$ and $P2'=\tau_{\text{air}}+\tau_{\text{sens}}$.

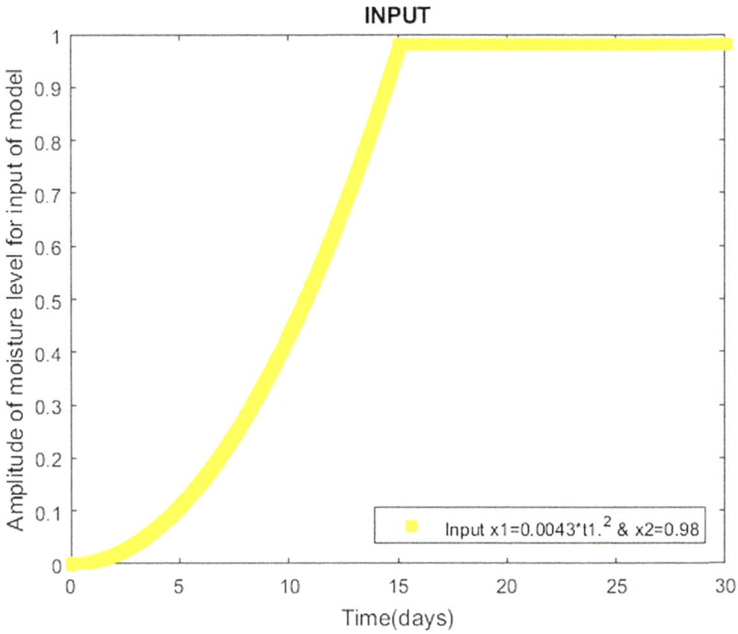


**FIGURE 7.7** Model indicating the level of moisture variation inside the breather.

The input signal is divided into two parts. As the first part of an exercise in the first 15 days, the nature of the signal is hyperbolic and it is indicated by

$$x_1(t) = 0.0043t^2 \tag{7.12}$$

During 15 to 30 ($15 < t < 30$) days, the strength of the signal remains constant, which is

$$x_2(t) = 0.98 \tag{7.13}$$

The output of the model in frequency domain for ($0 < t < 15$) is presented by

$$Y_1(s) = \left[\frac{(0.0087 + 0.130s)}{(900s^4 + 60s^3 + s^2)s}\right] \tag{7.14}$$

The output of the model for ($15 < t < 30$) is given by

$$\frac{Y_2(s)}{X_2(s)} = \left[\frac{(1 + 15s)}{(900s^2 + 60s + 1)}\right] \tag{7.15}$$

where $Y_2(s)$ is the output in the frequency domain for the input $X_2(s)$.

$$Y_2(s) = \frac{(1+15s)0.98}{(900s^2+60s+1)s} + \frac{900s^2y_1(15)+900s\dot{y_1}(15)+60sy_1(15)-15sx_2}{s(900s^2+60s+1)} \tag{7.16}$$

where the initial condition for $y_2(0) = y_1(15)$ and $\dot{y}_2(0) = \dot{y}_1(15)$.

## 7.2 Simulation response of model

As shown in Fig. 7.8, a model represented by Eq. (7.16), which was simulated the response along with the input signal, is demonstrated as the summation of an individual response $Y_1(s)$, $Y_2(s)$, respectively. It was simulated using MATLAB software. It indicates the response of the humidity sensor with time. The amplitude of the model response is smaller than the input signal due to pole positioning on the s-plane as far as the developed model is concerned. It consists of two poles, which are located at s = −0.03267 and s = −0.03398. It is considered to be highly stable as all the poles of the model fall in the left half of the s-plane. The response of the model was found satisfactory. The pole at s = −0.03267 is considered the dominant pole as it is very close to the imaginary axis. There is a derivative term in the numerator of the model which is why it has better rise time. The response of the model gets saturated after some time which indicates the saturation level of silica gel and steady-state

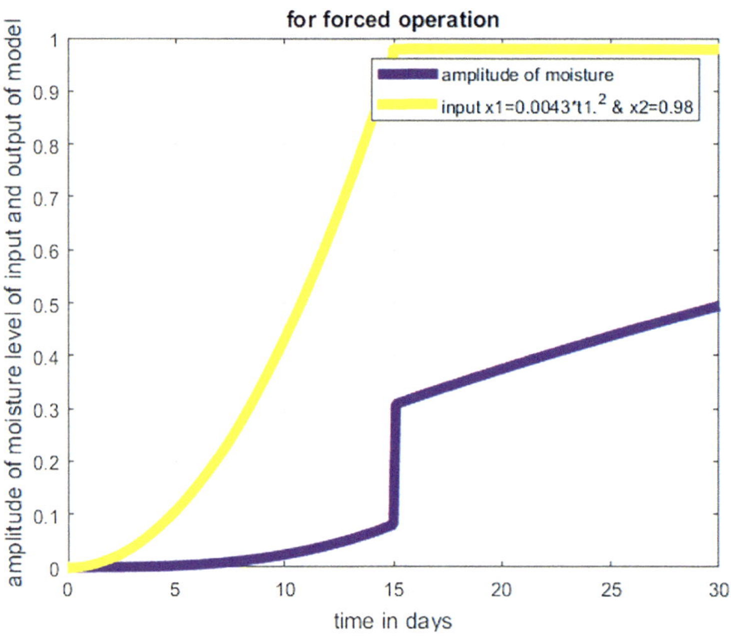


FIGURE 7.8 Simulation response along with the input signal.

response of the system. It may be the indication of the threshold level of water content inside the gel. It shows the saturation condition of the gel, and further, it cannot be useful for the breather condition until it is refreshed.

At this state, the response shows there is a need for dehydration of the gel or replacement time of the gel from the breather. The response of the model and input is shown in the common graph for better visualization. Model response is shown by blue colour and input by yellow colour.

## 8. Conclusions

This chapter presents all the sensors required for online monitoring of different parameters related to the health of a sub-station and in particular of a switchgear. These parameters can be broadly divided into two categories, namely, electrical parameters and physical parameters. In general, critical electrical parameters are voltage, current, power and power factor, while the physical parameters are temperature, gas leak and humidity. The role and impact of these parameters are briefly covered in this chapter. Furthermore, this chapter has extensively covered recent findings reported in research papers about the role and impact of moisture in the transformer.

Breather is an important part of high-voltage transformers. Condition monitoring of a breather is an integral part of the health monitoring of a transformer. The atmospheric moisture and dust particles ingress during the breathing process of a transformer degrade the life of the insulation system. A breather is used to protect the transformer from atmospheric moisture contamination and allow the dust-free air to enter inside the conservator during the period when it sucks atmospheric air during the decrease in pressure inside it. The method of installation of a breather in pictorial format is demonstrated in this chapter. It will be helpful for the new user in the proper installation of a breather to the transformer so that it can be utilized for optimal use for protection against excess moisture. Maintenance of the breather is very important for its efficient performance. Simulation of the developed model of a second-order differential equation is achieved by applying Laplace transform. Some important indices are as follows: (1) the breather should be free from the crack, (2) the colour of the silica-gel should be monitored regularly and (3) the level of oil inside the cup should be to the notch marked. Various parameters are used to decide the size of a breather. It consists of transformer rating, air space available at the conservator, the oil intake capacity of the transformer, the daily breathing rate of a transformer as well as its geographical location. The moisture content of silica-gel is the most important factor that decides the life of the transformer breather. The breather should be equipped with a programmable heater to regenerate its desiccant at a threshold value of moisture content that is allowable for transformer applications. Mathematical model-based moisture monitoring of the silica-gel is the recent trend to know its saturation level. Mathematical models indicating the amplitude of moisture content absorbed in silica-gel with time are developed. Two types of models, namely, transient sub-model and steady-state mode are developed. Breather air moisture content and temperature are considered as input to formulate the moisture estimating sub-model for the silica-gel.

## References

[1] S. Turrin, B. Deck, M. Egman, I. Cavalli, Medium voltage equipment monitoring and diagnostics: technological maturity makes concepts compatible with expectations, in: Proceedings of the 23rd International Conference on Electricity Distribution; Lyon, France, 2015.

[2] J. Liu, G.M. Huang, Z. Ma, Y. Geng, A novel smart high-voltage circuit breaker for smart grid applications, IEEE Trans. Smart Grid 2 (2) (June 2011) 254–264.

[3] M. Kamei, O. Takai, Influence of sensor information accuracy on condition-based maintenance strategy for GIS/GCB maintenance, IEEE Trans. Power Deliv. 26 (2) (April 2011) 625–631.

[4] IEEE Guide for Moisture Measurement and Control in SF6 Gas-Insulated Equipment, Standard C37.122.5-2013, December 2013, p. 142.

[5] S. Okabe, S. Kaneko, T. Minagawa, C. Nishida, Detecting characteristics of SF6 decomposed gas sensor for insulation diagnosis on gas insulated switchgears, IEEE Trans. Dielectr. Electr. Insul. 15 (1) (February 2008) 251–258.

[6] B. Bhalja, R.P. Maheshwari, N.G. Chothani, Types of circuit breaker and their testing, in: Protection and Switchgear, first ed., Oxford Univ. Press, New Delhi, India, 2013, pp. 373–376.

[7] M. Abdollah, A.A. Razi-Kazemi, Intelligent failure diagnosis for gas circuit breakers based on dynamic resistance measurements, IEEE Trans. Instrum. Meas. 68 (9) (September 2019) 3066–3077.

[8] Shaw Moisture Meter [U.K] Ltd, SADP, Datasheet, 2020. https://www.shawmeters.com/.

[9] T. Yuan, J. Shao, Z. Wei, Q. Wang, G. Wu, Y. Zhang, Exploration of the relationship between tyndall effect and moisture content in transformer oil, IEEE Trans. Dielectr. Electr. Insul. 30 (4) (August 2023) 1548–1556.

[10] S. Kumar, T. Islam, K.K. Raina, Modelling of breather for transformer health assessment, IET Sci. Meas. Technol. 11 (2) (March 2018) 194–203.

[11] P. Mishra, S. Kumar, T. Islam, An accurate model of breather for moisture estimation for transformer health monitoring, in: Lecture Notes in Electrical Engineering Book Series vol 886, Springer, Cham, June 2022, pp. 385–395.

[12] R. Nakane, H. Okubo, K. Kato, HVDC electrical insulation performance based on charge activity in oil-pressboard composite insulation structures, IEEE Trans. Dielectr. Electr. Insul. 26 (2) (March 2019) 576–583.

[13] Bharat Heavy Electricals Limited, Silica Gel Indicating Type, 2005, November 15. https://bhel.com/sites/default/files/Spec%20-%20AA55619%20%28Silica%20Gel%29.pdf.

[14] S. Kumar, L. Kumar, T. Islam, K.K. Raina, Condition monitoring of transformer breather using a capacitive moisture sensor, IEEE Trans. Ind. Electron. 67 (November 2020) 9779–9789.

[15] R. Emadifar, N.T. Kalantari, V. Behjat, R. Najjar, Monitoring and condition assessment of insulation paper of distribution transformers with novel oil spectroscopy method, IEEE Trans. Dielectr. Electr. Insul. 29 (5) (October 2022) 1904–1912.

[16] P. Milovac, R. Javora, V. Skendzic, Sensor technology in a medium-voltage switchgear for the US market applications, CIRED 2017 (1) (2017) 432–435.

[17] Z.H. Zargar, G.R. Biswal, T. Islam, A linear capacitive sensor for ppm moisture measurement in sf6 gas-insulated switchgear, IEEE Trans. Instrum. Meas. 70 (2021) 1557–9662.

[18] G.R. Biswal, T. Islam, S.D. Das, K.J. Akram, Ultrasonic Sensor Based Gas Density Monitoring of SF6 Gas Insulated Switchgear, Indian Patent No. 504038, January 29, 2024.

[19] H. Mingxia, et al., Research and design of automatic dehumidification device for transformer conservator oil desiccant, in: 7th International Conference on Information Science and Control Engineering (ICISCE), Changsha, China, 2020.

[20] A. Gil, P. Grange, Application of the Dubinin-Radushkevich and Dubinin-Astakhovequations in the characterization of microporous solids, Colloid Surf. Physicochem. Eng. Aspect 113 (1996) 39–50.

Chapter 8

# Incorporating communication protocols for substation automation

**Balakrishna Pamulaparthy[1] and David Macdonald[2]**
[1]*GE India, Hyderabad, Telangana, India;* [2]*GE Vernova, Cambridge, MA, United States*

## 1. Introduction of substation

A substation is an important part of the electricity generation, transmission and distribution value chain. It performs voltage conversion from high to low or low to high and supports several other important functions for control of the power. The main component of the substation includes power transformers. Substations are generally owned and operated by an electrical utility or may be owned by a large industry.

The following are the different types of substations [1]:

- Transmission substation
- Distribution substation
- Collector substation
- Converter substation
- Switching substation
- Traction substation
- Mobile substation

The following are the major elements of substation:

- Electrical power transformers
- Instrument transformers
- Conductors and insulators
- Isolators
- Bus bars

**Switchgear Design, Operation, and Maintenance Using Industry Standards**
**https://doi.org/10.1016/B978-0-443-24718-7.00008-1**

- Lightning arresters
- Circuit breakers
- Relays
- Capacitor banks and other miscellaneous equipment

## 2. Overview of substation automation

The increasing demand in consumption and savings of energy and the increased cost in bringing up new substations have given rise to the need for different utilities to manage their electrical systems more smartly and efficiently. As a result of this, the manufacturers have started exploring the requirements for a model of smarter substations with the aid of enhanced technical features and innovation, which could address the challenges enumerated in the following. On a broad way, the problem areas identified and the plausible solutions in the electrical substations for being refurbished into a smart one with automation are as follows:

- Reduction in copper volume and execution time of project by eliminating field labour into the factory. Incorporation of process bus based on IEC 61850 is the key enabler [2].
- Deduction in the data collection time to SCADA from the existing typical time of 1 second to 1 power cycle (order of 20 ms) thereby introducing an effective way of real time system. Introduction of synchro-phasors is the key enabler for this.
- There should be a universal communication protocol integrating all levels which is enabled by communication protocol IEC 61850 [3].
- For the access of data for supervision and decision-making at various levels, there must be an easy asset management system. For enabling this, standard database has to be prepared for SCADA (Supervisory Control and Data Acquisition System), EMS (Energy Management Systems), GIS (Geographical Information System), etc. [4].

## 3. Smart substation enabled with automation

A smart substation may be called as a sophisticated combination of different smart primary high voltage equipment and several other secondary devices, which are hierarchically placed in a proper sequence. Standardized communication protocols such as IEC 61850 enable interoperability between the devices and proper sharing of vital information in smart substations. This protocol is an important basis for the formation of smart substations and a path breaking technology for future smart grid. Modern smart substation design parameters should take into account high and specific requirements for security, reliability, interoperability, controllability, maintainability, re-configurability, reduced cost, flexibility and impact on the environment. Modern substation design must also take into account higher requirements on reliability, security, interoperability, re-configurability, controllability, maintainability, flexibility, reduced cost and environmental impact. In IEC 61850, the sampled values from the field instrument transformers are sent to the process bus and the relays for control and protection. The stability also improves with the replacement of the conventional instrument transformers by non-conventional instrument transformers (NCITs) [5]. For the different intelligent functionalities to be implemented, there should be standardized protocols for an integrated system architecture. The important part of smart substation communication is to dynamically update and share the information from the primary equipment and the control centre. So the preliminary functions of the smart substation include (1) obtaining and sharing real-time data from the field equipment, (2) interaction between the different devices for information exchange for proper dispatch and maintenance, (3) assessment of life cycle of assets based on the data representing the real time conditions and (4) supporting power system control, smart regulation and other much advanced applications.

## 4. Automation technology

Within the boundaries of a smart substation with automation, the popularly known three-layer two network structure is followed where there are three physical layers connected by two network layers are included. The IEC 61850 plays an important role in the implementation of object-oriented communication which supports handy multivendor solutions [6].

The smart high-voltage primary equipment are equipped with sensors and actuators which ensures flexible sharing of digitized data between primary and secondary equipment and so a comprehensive monitoring of high-voltage equipment is possible. The information from the smart NCITs is digitized fully; as a result, there is reduction in huge volume of hardwired copper cabling by high-speed fibre optic cables. This ensures a uniform interface and reduction in carbon footprint as depicted in Fig. 8.1. The essential difference between the conventional substations and smart substations is that the former functions based on original engineering design whereas the latter has functionalities based on field configuration and modular structure. The characteristics and features of a typical smart substation can therefore be enumerated as follows:

1. Customized and tailor-made functionalities based on architecture, framework and design.
2. Integrated, digitalized secondary system and smart, modern primary equipment.
3. Digitized data, standardized model, networked information exchange, integrated applications and CBM (condition-based monitoring).
4. Increase in the controllability and observability of the system.

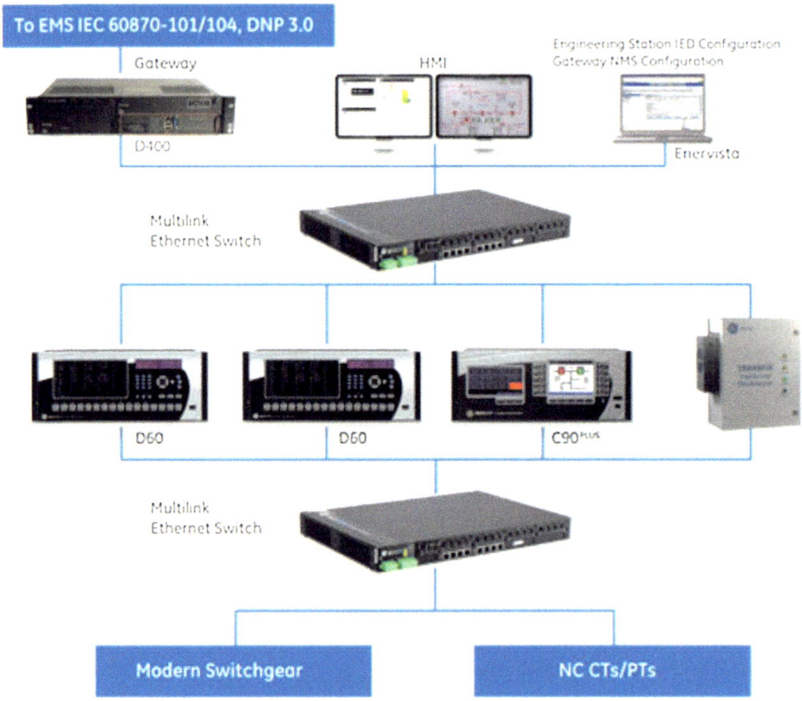


**FIGURE 8.1** Substation automation system.

5. Gradual shift from traditional ways of 'passive regulation' to 'active control'.
6. Enhanced and improved security, flexibility, and reliability, and more environmentally friendly and compatible carbon footprint.

The changes affected in smart substations due to the evolution into smart one are many. The main aspects to be affected by such changes are discussed in the following.

## 5. Intelligent equipment for substation automation

The characteristics and features of smart and modern high-voltage equipment in a smart substation are that they are capable of networked control, digitized measurement, integrated functionality, monitoring of states and exchange of interactive information. It is implemented by intelligent electronic devices (IEDs) embedded for monitoring and control, measurement and protection of the host equipment. Two of the most important high-voltage equipment are NCITs (nonconventional instrument transformers) and SSDs (smart switching devices).

The NCITs have done away with the problem of electromagnetic saturation by replacing the conventional ferrous core of the current and voltage transformers. NCITs in general have the advantages such as smaller and compact size, light weight, reduced price, higher accuracy, longer life span, broader bandwidth and better, improved transient performance. The accurate information gathered by the NCITs can be exchanged and circulated for better decision-making and control actions across the bay control as well as station control in multicast way.

SSDs are another group of key enablers for smart substation with automation [7]. Varying from conventional switching devices, SSDs adopt the standard communication interface (IEC61850) and integrate the functions of the measurement, condition monitoring and control as well as protection and communication, satisfying the necessary conditions for integrated and continuous power flow, service business flow and information flow. An example of a smart high-voltage device is an HV transformer embedded with sensors to measure oil and winding temperature, on-load tap changing position, partial discharge condition, etc. Another important aspect of smart HV equipment is the penetration of power-electronics-based solid-state transformers [8] which helps in faster response time in the event of faults. Faster information processing systems are needed to enable this. The condition-based maintenance and better utilization of life cycle of the equipment is possible due to better integrated design and increased controllability.

## 6. Smart substation with automation – Driving factors

Presently, almost all the components in a substation are enriched and augmented by some kind of technology which is based on embedded systems thereby enabling them to be operated at higher value of ratings with greater degree of reliability than was done hitherto. The challenge presented now involves conglomerating all these elements into a totally integrated and digitalized substation and making it work in a suitable environment. The key changes that can be identified from an analog to a digital substation include the following:

- In the substations digitalization of the analog devices.
- The increased demand of power consumption at the consumer end with enhanced power quality.
- The need of reduction in the fault clearing time.
- Internet of things (IoTs) [9] have replaced conventional components in the customer side which is testament to the penetration of the digital technology. IoT may be defined as a networked mesh of devices, vehicles, buildings and almost all types of physical object one can perceive that are connected by networks, equipped with sensors, and are monitored and remotely controlled by computer aided systems.

## 7. Smart is not a new concept

The transmission and distribution systems today is the aftermath of the substations enabled with technology with the stratagem of an all-knowing grid written all over it. The naming used might have been different, but the idea of smart substation was very much there. It was just a matter of time for the technology to get mature. The SCADA was deployed by the industry, and then the introduction of remote terminal units (RTUs) took place. SCADA is a type of architecture for control system that makes use of computers, data communications and graphical over network and human–machine user interfaces for a high-level of process supervisory management. It uses many other peripheral devices such as programmable logic controllers (PLCs) and several discrete PID controllers to interface between the process plant and the machinery [10]. Then the introduction of microprocessor-based relays was incorporated into the protection systems. With the technological advancements, communication interfaces were directly integrated into the microprocessor-based relays. Further enhancements with microprocessors' modular hardware and firmware (software) design, the introduction of multifunctional microprocessor-based relays took place. This helped relays to directly communicate with RTUs and gateways in substation. Incorporating multiple and different microprocessor-based relays into the electrical system operation throughout the substation is the real challenge.

## 8. Standards are the key

The initial problem that cropped up with the introduction of IEDs was that they were proprietary equipment with proprietary software and communication facilities. Open standards became the solution as one manufacturer's equipment and software did not fit into the software of another manufacturer's firmware. The investigation and work on IEC 61850 by the International Electro Technical Commission (IEC) started in 1995 and published as an accepted communications standard in the year 2003 for automation systems in electrical substations [11]. The documentation of IEC 61850 has several parts that pertain to modelling of data, reporting schemes of information, fast transfer of events, transfer of sampled data, control of and communication among substations. World's first IEC 61850-based substation came up in Switzerland by the year 2005.

## 9. The digital divide

The electrical industry system is a hybrid between the digital and the analog technologies. The definition of the digital substation is constantly varying because technology is changing so rapidly. The popularly used and established communication technology is that of the high-speed bidirectional and multipoint communication systems. Digitalization of substation devices enables them to communicate directly with each other over digital substation on the same network. The communication protocol, namely, generic object-oriented substation events (GOOSE) protocol as defined by IEC 61850 has facilitated this, intended to supplant the traditional hardwired communications gradually (i.e. all the metallic contacts and copper wiring) by fibre optic and ethernet cables between the IEDs.

## 10. Digital measurements

The fibre-optic current and voltage sensors generate much faster responses and, thus, result in much precise measurement as compared with the conventional magnetic-core-based CTs with the help of a digital interface, which are essential for integration and incorporation into the digital substation. The fibre-optic current and voltage sensors may be classified as a group of transducers connoted as non-conventional instrument transformers (NCITs). They have the features to provide improved safety by insulating the secondary equipment from high voltages of the primary side by means of fibre-optic cables rather than only copper wire. This leads to decrement in the substation's carbon footprint and enables more compact and concise designs.

## 11. Global communication standard for substation automation (IEC 61850)

**Why IEC 61850 – A Revolution in Substation Communications and Automation?** Communication is an important aspect of power system for control and stable operation. In conventional substations, the communication between the physical devices such as CTs and PTs was by the means of physical hardwired copper cables and in similar way relay issued trip signals to the breakers. Thus, each device needs to be physically connected by hardwires to every other device on the network so as to enable exchange and flow of information. With the advent of ethernet communication, the physical wires were replaced by faster fibre-optic cables; thereby a major shift in communication technology was achieved. With the introduction of IEDs and phasor measurement units, their integration needed a common protocol. The IEDs communicating with each other via GOOSE protocol would pose difficulty for refurbishment and interoperability. IEC 61850 is that open standardized communication protocol for power system utilities that provides interoperability between IEDs of different vendors and defines a common set of rules for substation automation system (SAS) by defining logical nodes. Apart from standardizing the different components of the electrical utility, the 61850 protocol also is commensurate with 'self-description' i.e. ability of a device to make its data available to a client browsing the contents of the standard. The standard describes various aspects for client—server and peer-to-peer communications, substation design and configuration, testing and project implementation.

The IEC 61850 is an extensive suite of standards, composed of over 27 documents and 4000 pages, making up 10 parts [11], which are enlisted in Table 8.1.

Earlier the protocols used to dictate how the bytes of data would be transferred across the different devices. However, IEC 61850 models device in such a way as to dictate how the components of the power system should organize their data so that their data is visible across all manufacturers. The modelling approach of devices [12] as per IEC 61850 consists of (1) physical device, (2) logical device, (3) logical node and (4) data class:

Each physical device such as relays consists of several logical devices, which in turn consist of several logical nodes such as circuit breakers and

**TABLE 8.1** IEC 61850 sections and definitions.

| IEC 61850 sections | Definitions |
|---|---|
| IEC 61850-1 | Introduction and overview |
| IEC 61850-2 | Glossary |
| IEC 61850-3 | General requirements |
| IEC 61850-4 | System and project management |
| IEC 61850-5 | Communication requirements |
| IEC 61850-6 | Configuration description language for substation IEDs |
| IEC 61850-7-1 | Basic communication structure for substations and feeder equipment – principles and models |
| IEC 61850-7-2 | Basic communication structure for substations and feeder equipment – abstract communication service interface |
| IEC 61850-7-3 | Basic communication structure for substations and feeder equipment – common data classes |
| IEC 61850-7-4 | Basic communication structure for substations and feeder equipment – compatible logical node (LN) classes and data classes |
| IEC 61850-8-1 | SCSMs – specific communications service mappings (MMS) |
| IEC 61850-9-1 | SCSM – sampled values over serial multidrop point-to-point link |
| IEC 61850-9-2 | SCSM – sampled values over ISO/IEC 8802-3 |
| IEC 61850-10 | Conformance testing |

transformers. They contain several data classes which are analogous to files stored in computers. Suppose there is a logical device, relay 1 that comprises of a single circuit breaker logical node XCBR 1 for which it is required to know whether it is in the remote or local mode of operation. Based on this modelling hierarchy, an object in IEC 61850-based communication hierarchy is named as shown in Fig. 8.2.

## 12. Logical nodes

A Logical node (LN) is a named grouping of data and associated services that is logically related to some power system function as shown in Fig. 8.3. The real devices such as breaker, transformer, isolator, etc. are represented as logical nodes in virtual world (example: breaker is shown as XCBR). Once the logical node is created, it comes with standard data objects representing breaker measured data and configuration file for breaker operation modes. The standard data from each logical node is mapped over Manufacturing Message Specification (MMS) protocol to transfer data to control centre over TCP/IP.

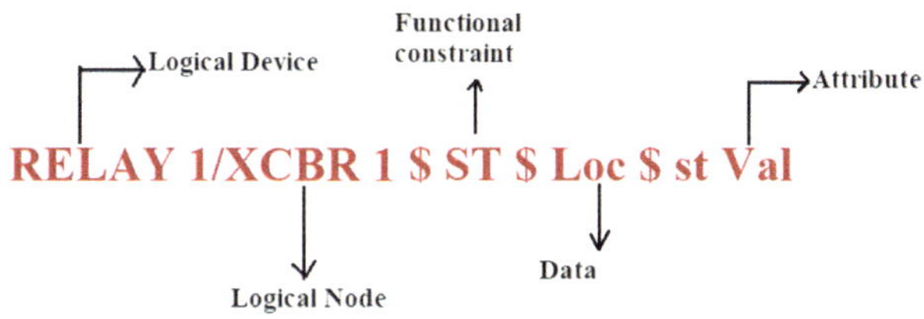


FIGURE 8.2 Enumeration of IEC 61850 logical node.

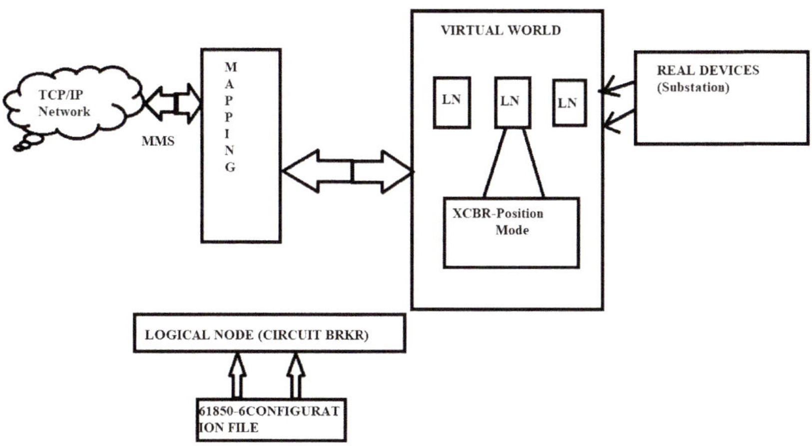


FIGURE 8.3 Logical nodes.

## 13. Overview of the IEC 61850-based communication

IEC 61850 is not just a set of rules and procedures for communication like Modbus nor DNP 3. While it is true that IEC 61850 defines communication services and protocol stack, it also includes the following:

- Data model
- Configuration language
- Configuration file types
- Engineering processes

This combination provides interoperability between devices of different manufacturers:

- The data model creates an abstraction of the electrical network and allows it to be represented with semantically meaningful addresses and data objects, which can be understood by the devices of other manufacturers.
- The configuration language and file definition are what allow third-party tools to do the engineering, the preparation of the configuration files for all the devices with a substation.

Not only does IEC 61850 offer the advantage of interoperability, but it also makes a top-down engineering process possible. This is what guarantees the compatibility between different devices within the same Substation Automation System (SAS) and thus reduces much of the cost in preparing configuration files (engineering) and in testing during Factory and Site Acceptance Testing.

In this example of a single line diagram of a typical substation shown in Fig. 8.4, we can see that we have the substation HMI at the top, the different IEDs at the bay level and then a merging unit, or process interface unit, which digitalizes the measurements from the current and voltage transformers in each

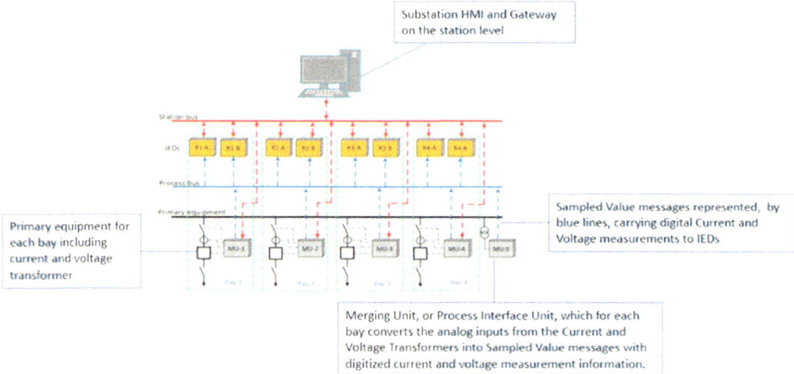


**FIGURE 8.4** Substation automation system.

bay position of switchgear and handles the digital inputs and outputs of the switchgear.

These measurements are sent to the IEDs, which can process them all or downsample these inputs to process a fraction of the samples per cycle. Harmonic filters are applied, and these samples are converted into phasors and input into logical algorithms, which determine whether any output action should be taken, such as a trip command or alarm signal. These devices are normally multifunctional and in one device include several functions such as protection, control and communication [13]. These control functionalities include the sending of interlocking commands to adjacent IEDs. Medium-voltage bay positions usually contain just one multifunctional IED with all the protection, measurement and control functionality needed for that bay. High and extra high-voltage positions quite often have one IED, which includes the control functionality and a simple protection function such as overcurrent protection and other separate IEDs, which cover the distance and line differential protection. This is due to the dependencies between line distance and differential protections on each extreme of the same line: they must be from the same manufacturer and model.

All these IEDs act as servers, which is to say that it provides data to and receives commands from the client devices. Typically, there are two types of client devices interfacing with the IEDs: the substation human–machine interface (HMI) and the substation gateway. The HMI is principally employed for realizing local switchgear operations and helping with diagnosis of device, communication and switching failures. The substation HMI allows for localized switching operations that realized a safe distance from the actual switchgear, which is especially relevant for earthing operations, which should never be realized from the remote-control centre. Switching from the substation HMI is normally done from a single line diagram viewer of part or all of the substation, where the switchgear can be selected and operated. Before this, however, the operator must change the level of control authority so that operations cannot be taken simultaneously from the remote-control centre. All the devices within the SAS are self-monitoring, which means that they can send device or communication failure alarms to the client devices. Therefore, the substation HMI also contains pages for the operator to view and recognize the alarms sent by the IEDs.

The substation gateway is principally a protocol converter, translating signals from IEC 61850 to telecontrol protocols such as IEC 60870-5-104 and translating commands in the opposite direction.

Given that this device is what allows for remote monitoring and control of the substation, it is the most critical device in the substation and as such is often duplicated with a redundant device in active-standby mode.

These devices can communicate across the substation local area network (LAN) thanks to the ethernet switch. This switch, otherwise called bridge device, is what makes ethernet deterministic and collision free, all messages are sent via the switch, which has a small buffer, in which messages form a queue before being forwarded onwards according to their destination MAC addresses.

The station bus is displayed in red, and the process bus is displayed in Fig. 8.4.

- The station bus allows us to replace the point-to-point hardwired connections, both horizontally between IEDs and vertically between bay-level IEDs and station-level devices such as HMI or gateway.
- The process bus is for sending time-synchronized high-priority sampled value messages from the process interface units to the IEDs. These are basically measurement information, current, and voltage values and are of course sent in real time.

All these signal connections must be programmed and stored into each device as part of the engineering process. This is realized by two main types of tools, the ICT and the SCT.

The **ICT, IED Configuration Tool**, is the manufacturer-specific software tool; it is the tool that loads the final configuration to the IED device and can be used for interactions with the device throughout the lifecycle such as settings changes. For example, where after a protection mis-operation and recalculation of settings, an engineer could employ the ICT tool to communicate with the IED and apply a new overcurrent setting value.

The **SCT, System Configuration Tool**, is a third-party tool, which creates the complete IEC 61850 defined configuration for the whole substation. It does this by inputting the IED Configuration Description (ICD) template files and by drawing the single line diagram and creating or importing the System Specification Description (SSD), a file type that is described in the following paragraph.

These tools import, produce or export the following standardized file types in the engineering process.

- The **ICD, IED Configuration Description.** This is the device template that contains the device data model as well as the client and server capabilities of the device, for example how many of each type of messages it can send or receive. This self-description allows even third-party configuration tools to know the limitations of each device when preparing the IEC 61850 configuration of the substation.
- The **SSD, System Specification Description.** It is like a functional specification of the protection and control requirements of the substation. It is often created by drawing a single line diagram and having the SCT tool convert this to SCL language. Simply put, the SSD file is a description using IEC 61850s own language, of both the primary electrical systems: voltage level, bay position, switchgear type, and the Protection and Control Requirements within each bay.

- All the ICDs, IED Configuration Descriptions and the **SSD, System Specification Description** of the substation are imported into the SCT to create the System Configuration Description (SCD) file. This is the IEC 61850 configurations for the whole substation.
- This **SCD** is then imported by the device manufacturers IED Configuration Tool to create the CID file, the Configured IED Description, which is the configuration file which is finally loaded to the IED.

While part 6 of IEC 61850 offers both bottom-up and top-down engineering processes, it is the latter which has had most impact. A top-down process is possible because a third-party configuration tool can be employed regardless of the manufacturer of each device. A top-down process as shown in Fig. 8.5 inherently provides consistency between the different devices within the same SAS, reduces the amount of effort within the configuration process and reduces the requirements for Site Acceptance Testing. The top-down process produces a coherent configuration file (SCD) for the whole substation, which can be used as an input for simulation and testing devices.

Before IEC 61850, information within an IED was structured using numerical lists, and for this, an additional signal list was required to understand the electrical significance of each number. Therefore, any interoperability that was achieved was through a slow and error-prone process. The use of IEC 61850 changed this because it provided a common data model. The data model provides an abstract and standard way to describe and organize all the data regarding the protection and control system as well as the physical electrical system assets. In IEC 61850, the model is organized in a hierarchical fashion so that there is a unique identification for each function within the system. These functions are called logical nodes. In this example of gas-insulated switchgear shown in Fig. 8.6, there are logical nodes to describe the physical electrical network:

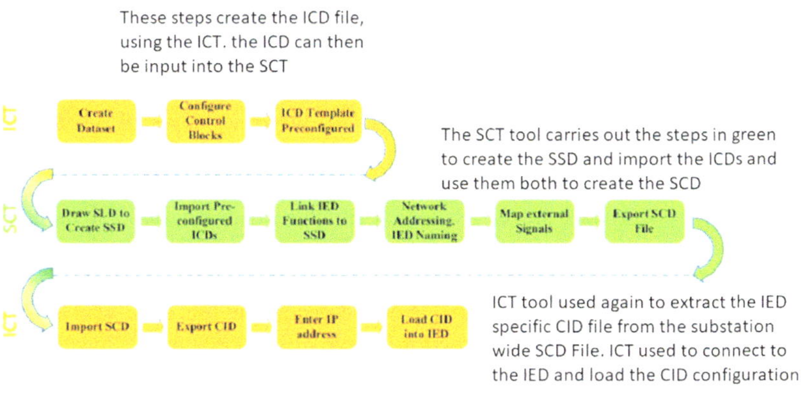


**FIGURE 8.5** Flowchart of engineering process.

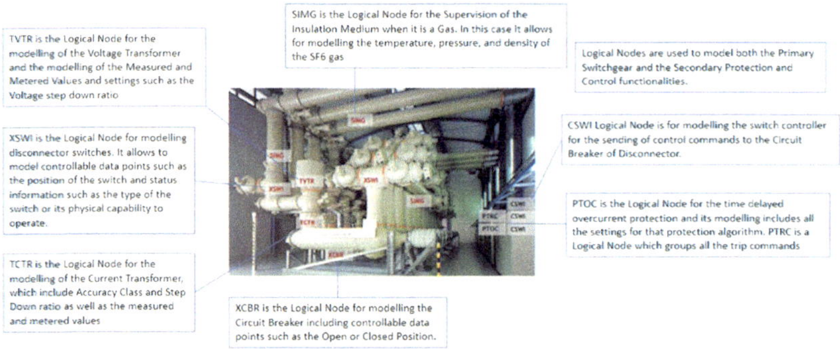


**FIGURE 8.6** Typical picture showing the primary and secondary system logical nodes within a GIS substation.

- LN SIMG Gas Insulation Supervision
- LN XSWI Circuit Switch
- LN TCTR Current Transformer
- LN TVTR Voltage Transformer
- LN XCBR Circuit Breaker

And to describe the Protection and Control System:

- LN PTRC Protection Trip Conditioning
- LN PTOC Time overcurrent Protection
- LN CSWI Switch Controller

These functions are included also in other devices and bays, so we need to identify the device where they are implemented, so that each functionality in each device can have a unique address. Therefore, to identify the IED, we have the logical device name. This logical device name is usually made up of substation name, voltage level and bay name. Take, for example, a logical node for a circuit breaker, there are many signals we could have for the breaker: its position, its local/remote status, interlocking status and all these types of signals are called data objects within IEC 61850. For one position, there are multiple attributes that we could have such as the value of that position, whether the breaker is open or close, the timestamps of when it last changed and the quality of that information (for example, whether the signal is valid or not such as when the value is oscillatory, or out of range, or inaccurate). These signals are called data attributes. This hierarchical model of data manages to give all the signals a unique address that is semantically meaningful and understood by devices from other manufacturers [14].

In the example PIOC Logical Node depicted in Fig. 8.7, which is for Instantaneous Over Current Protection, this logical node should allow operator to enable or disable the protection element. It should also help to set the pickup value (threshold) at which it should trip and allow for specific output signals to

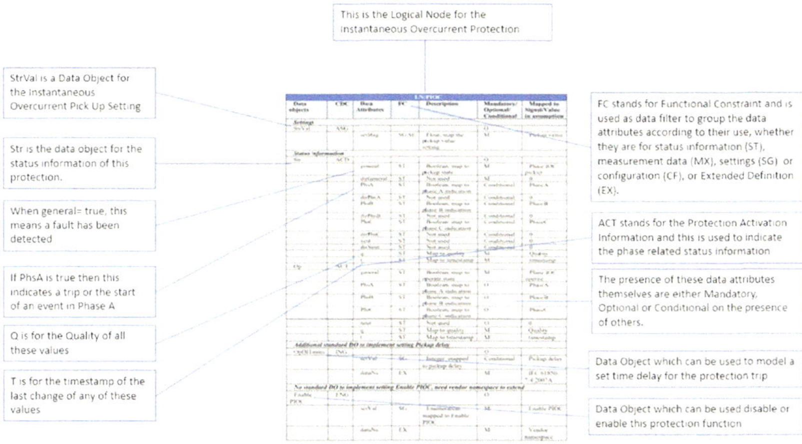


**FIGURE 8.7** Logical node for instantaneous overcurrent protection.

let operator know the elements that may have tripped, i.e., which phase has detected the fault. Therefore, we can see a data object STr Val for setting the pickup value, another for showing which phases have detected faults, adding a time delay, another for enabling/disabling the protection.

The example of a breaker failure application [15] can give a good insight into how the different logical nodes are interdependent on each other to build an application, even across logical nodes in different devices. So, this breaker failure works in such a way that once the protection for the faulted bay has already tripped, in the worst-case scenario that the breaker does not operate and the protection continues to see fault current, then it will need to send a trip signal to the other adjacent bays to stop them feeding the fault.

So, we can see from Fig. 8.8 that within the IED of the faulted line, the PIOC logical node whose trip signal is aggregated by the PTRC logical node, which sends the trip signal to the circuit breaker, the XCBR. It also notifies the breaker failure logical node that it has operated, which will then monitor the circuit breaker status to see if it opens, send it a re-trip signal, and then once the appropriate time delay has transpired, it will send a GOOSE message which is a peer-to-peer broadcast instantaneous message which is processed by other IEDs which have subscribed to it during the configuration stage. GOOSE message with trip signal to the adjacent IEDs so that their breakers can clear the fault. Of course, this activity needs to be logged and supervised so these green dotted lines represent the Report messages. The Report messages could be thought of as a vertical communication where the server IEDs would notify the Client Substation Gateway and HMI devices to register that the protection has tripped, that the breaker has not opened, that the breaker failure was activated and finally to register the trips and openings of the adjacent breakers.

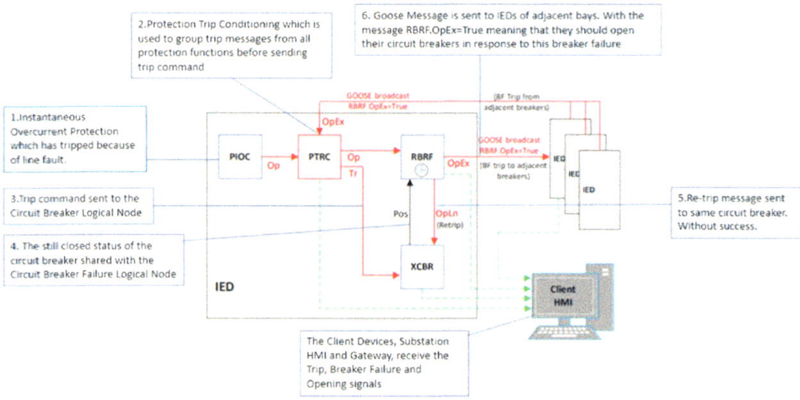


**FIGURE 8.8** Communication exchange between internal and external logical nodes for breaker failure application.

## 14. Station bus communication for switchgear operation and control

IEC 61850 has leveraged many of the advancements in communications technology, such as switched ethernet, transmission control protocol/Internet protocol (TCP/IP) and abundant bandwidth. IEC 61850 recommends the use of ethernet part of the LAN protocol. A local area network (LAN) is simply a group of interconnected devices which share resources. Unlike with point-to-point cabling, with a LAN, the path is not simply defined by the physical connection and thus each IED requires a unique identifier.

Initially, ethernet was compatible only with proprietary protocols and the interoperability remained elusive. The open system interconnection (OSI) and latterly the TCP/IP protocol stacks provide the rules and procedures so that different devices, from different manufacturers, can regulate their communication together. IEC 61850 communication services are based on the OSI and TCP/IP protocol stacks. For example, they state how possible problems should be resolved. These problems could be the loss of messages or receiver saturations (when the receiving node is overloaded and no longer able to clearly receive the bits being transmitted).

The OSI conceptual framework is made up of seven layers. The application layer defines the format of the message and presents the final information to the user. These are the highest-level protocols such as Manufacturing Message Specification (MMS) used for the sending of the Report messages, or IEEE 1588 Precision Time Protocol (PTP), used for distributing time synchronization messages to all the devices in SAS. The presentation layer does the encrypting and decrypting, it takes the messages from the application layer and transforms them into 1s and 0s and vice versa. The session layer is used to control the dialogue between the devices, from the authentication, maintenance, restoration and termination of the session.

The transport layer has a very important role, it adds a sequence number and error check; thus the IEDs can detect errors, and what happens when the errors are detected depends on the transport protocol. Transport control protocol (TCP) is a connection-orientated protocol, and so when errors are detected, then retransmission of the messages takes place. User datagram protocol (UDP) is not a connection-orientated protocol; it is uni-directional, and when errors are detected, then no retransmission of the lost message takes place. Therefore, TCP is suitable for messages which must certainly be collected by the receiving device, whereas UDP is most appropriate for messages where time is of the essence and messages would become obsolete before retransmission.

The network layer, or layer 3, is what manages communications where there is not a physical connection. This means that it does the addressing and routing. The Internet protocol (IP) address should be used to assign network addresses for each IED and to determine the route that is used to the next device. Only messages that travel out with the substation have a network layer.

The data link layer, the layer 2 communication, is used for communication within the same LAN, within the same substation. It adds the link addresses and control information in the header; it adds the frame check sequence. It uses IEEE 802 Media Access Control and 802.2 Logical Link Control, and therefore, Media Access Control (MAC) frames and MAC addresses are utilized. Internal Substation Messages are said to be layer 2 messages because they do not travel outside the substation. No IP address is required, and instead, this communication can be realized using only MAC addresses.

The station bus connects all the devices, which make up the SAS and its principal objective is to provide monitoring and management and a communication link between the control centre and the IEDs. It facilitates both the multicasting traffic to many IEDs as well as the unicast traffic between the IED as a Server and a client such as the substation human–machine interface (HMI) or gateway. It usually consists of station-level control from a gateway or a substation HMI, a time synchronization system that could be simple network time protocol (SNTP) or in more recent cases precision time protocol (PTP), and the IEDs in each bay and a communication system. The station bus is used to carry a variety of services including the following:

- Multicast GOOSE messages between IEDs
- Unicast MMS between station-level devices and bay-level IEDs. This is used for the Report messages or the transmission of control commands to IEDs.
- Unicast time synchronization services such as simple network time protocol (SNTP)

- Other non-IEC 61850-specific protocols such as secure file transfer protocol (SFTP). This could be used for downloading COMTRADE oscillograph files and other fault information from the IED.

The GOOSE messages replace the hardwired connections between IEDs, and as such, the requirement for an IEC 61850 equivalent was that it could either deliver equal or improved latencies on what was then the fastest auxiliary relay, which was then 4 ms. Ethernet point-to-point connections could not deliver this performance for the worst-case scenario of a substation with 100 IEDs and so the publish subscribe model was defined for GOOSE messages [16]. In a publish subscribe model, the sender publishes its message without knowing beforehand who the subscribers are. Layer 2 GOOSE messages are horizontal messages broadcast within the LAN between IEDs. These messages are typically used for interlocking communication, breaker failure schemes or sending trip messages to the Merging or Process Interface Units (PIUs). The receiving device must currently check the Destination MAC (Media Access Control) Address within the message header to determine whether it is an accepted message to which it is subscribed. The destination MAC address for GOOSE and for Sampled Values is a fictitious address included within the Ethernet Header and within the configuration of each subscribing device. It is not a real hardware MAC and therefore the Switch is not able to automatically determine where the messages should be sent. That knowledge only comes from the SCD file. These layer 2 messages are therefore broadcast to all devices (except the sending IED), unless the switches are configured to filter for example using Virtual Local Area Networks (VLANs) which could be created, for example, for each GOOSE message thus reducing the unnecessary traffic flow. For the sake of speed, these GOOSE messages are mapped directly onto the ethernet frame as shown in Fig. 8.9.

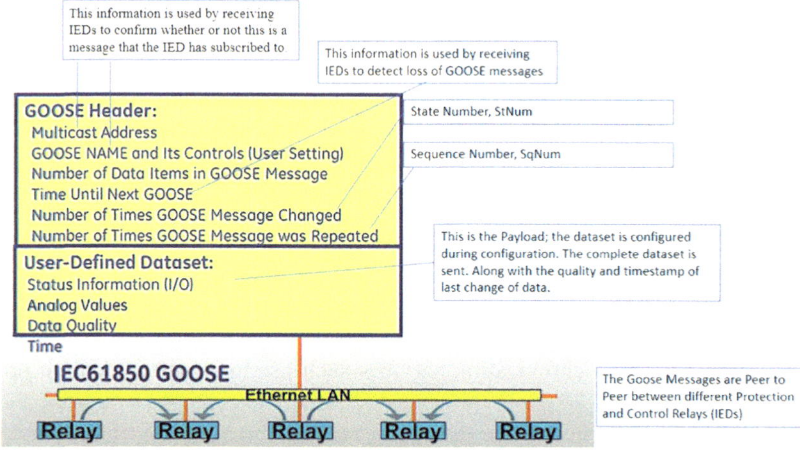


**FIGURE 8.9** Creation of GOOSE message and its content.

This means that there is no transport layer, there is no session layer, nor is there a network layer. The publishing device is not even aware of which are the devices that have subscribed. This means that the sending message receives no reception acknowledgement from the receiving end device.

To compensate for this lack of acknowledgement, GOOSE messages implement repetition and a Time Allowed to Live (TATL) mechanism. GOOSE messages are sent instantly upon a data change, but rather than wait until the next data change, these messages are periodically transmitted upon ever-increasing intervals until they have reached a stable situation where the messages are repeated following the maximum interval. This interval is called the heartbeat time and is typically less than 60 s. Inside every GOOSE message is a TATL parameter which informs the subscribing device of the maximum time that it should wait to receive a GOOSE message. If a GOOSE message is not received within this period of time, then the subscribing device is informed that the subscription has been lost or there is a communication failure within the LAN. This TATL parameter should be set to a value superior to two times the heartbeat time of the same GOOSE message.

GOOSE messages also contain some other key parameters such as Sequence Number and State Number. The State Number, StNum, is used to record the state changes within the GOOSE message payload. The Sequence Number, SqNum, is used to record the number of frames sent of the exact same GOOSE message. When there is a change in the GOOSE message payload, then the StNum increments by one and the SqNum starts again at 1. When a GOOSE message is simply retransmitted without a change in the data, then the StNum stays the same, and the SqNum increments by one.

These parameters allow the subscribing device to confirm the loss of messages in the case that a SqNum has been missed, or in the case that the StNum has incremented, then the subscribing device knows to check the new message payload content. While GOOSE messages are useful for the fast communication between IEDs, the vertical point-to-point communication between IEDs and client devices does not require so much speed but on the other hand does require reliability mechanisms, error checks and in some cases the resending of a message which has not been received properly. IEC 61850 provides Report messages for this. Report messages are only sent as part of a two-party association between the MMS client and MMS server. This two-party association is created via a three-way handshake. The client first sets up a connection from a random port, then the transport session is established and finally the MMS protocol will set up an association. This last part is conditional on both devices knowing each other's data models and service capabilities. This client server communication is used principally for both Reporting and for Control Services.

Two types of reporting models are defined in IEC 61850-8-1: Buffered Reports and Unbuffered Reports. Similar to GOOSE messages, these Reports have associated datasets, ordered sets of data attributes and data objects for sharing with the client device. Unlike GOOSE messages, where the complete dataset is always transmitted, for Report messages often only the values of the datasets are sent. Buffered Reports, as the name suggests, store the message within a buffer within the server during the message exchange, should the communication be interrupted, the message can be resent either in its entirety or uniquely those data attributes which were not received during the first attempt (this latter option is dependent on the EntryID parameter being present within the Report Control Block). These Buffered Reports are therefore appropriate for mission critical information which must be transmitted to the client, for example the alarms and status signals. Unbuffered Reports are not re-sent in the case of problems during the first attempt. This approach is therefore appropriate for time sensitive messages such as measurements which would become obsolete soon after publishing and are therefore not suitable for re-sending to the client devices, which will receive newer measurements reports anyway.

To configure the Report messages, there are two main tasks. The dataset must be created as shown in Fig. 8.10 with the ordered selection of all the data objects and data attributes that will be shared, then a control block should be configured which will refer to said dataset, state the message type whether buffered or unbuffered. When the message is initially being configured, the Configuration Revision should be set to 1. This parameter will increment every time a change is made to the control block. The optional fields need to be selected for inclusion into the message, for example if the sequence number is selected, then this offers a great diagnostic tool for detecting lost messages. The Report Time Stamp is of course highly relevant for the sending of measurement dataset values. If the Reason for Inclusion field is included as shown in Fig. 8.11, then the Report will be sent with the reason why this message was

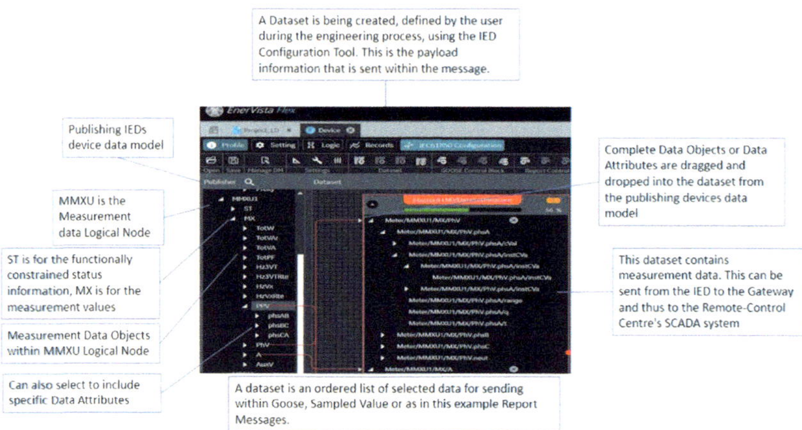


**FIGURE 8.10** Creation of a dataset for an unbuffered report.

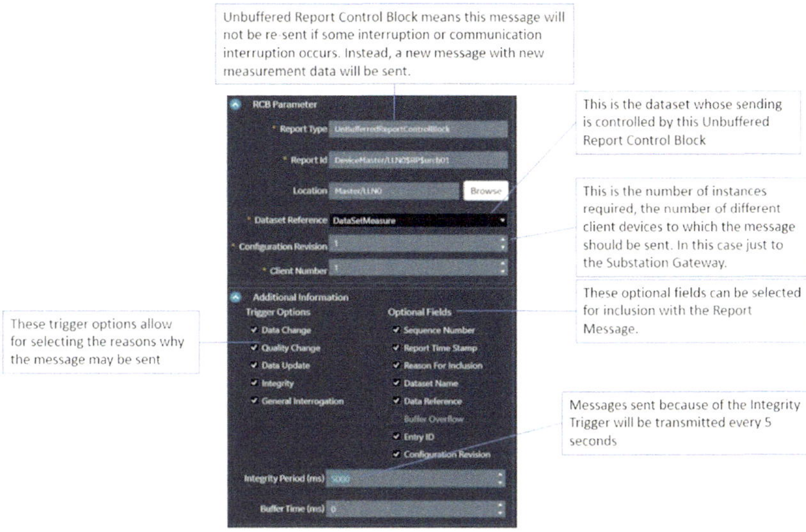


**FIGURE 8.11** Creation of unbuffered report control block.

triggered. Unlike GOOSE messages, Reports are not sent continually without a special reason. They need to trigger by a trigger option that has been a preselected reason for sending the message. For example, if Data Change is selected, then when the data actually changes, a Report will be sent. This would make perfect sense for the Position state of a circuit breaker, but measurement data should be changing continually. An appropriate trigger in this case would be the interrogation period, which is like a periodic polling so that every 5000 ms (in this example) the measurement data can be updated. Data Update means that the message will be sent even when a new measurement is taken with the same values. General Interrogation allows for Reporting upon the specific request of the client. In the case that this same Report be sent to both the Substation Gateway as well as the Substation HMI, then at least two instances should be created.

## 15. Switchgear control using IEC 61850

The IEC 61850 data model defines the logical nodes, data objects and data attributes not just for protection functions but also for the switchgear and their associated controls. These data models can be interrogated, and the data attributes changed via the MMS protocol. The IEC 61850 defines in part 7−2 control models so that client devices can send control commands to Switchgear via Bay Control Unit IEDs. Four state machines are defined for the sending of the controls:

- Select Before Operate
- Select Before Operate with Enhanced Security
- Direct Operate
- Direct Operate with Enhanced Security

The Direct Operate allows for commands to be written directly to the hardware, whereas the Select Before Operate requires that two commands be sent and requires a response from the Bay Control Unit so that the Switchgear is reserved for the client before the command can be executed. The Enhanced Security options are not, as could be suggested from the title, about cyber security, but rather about giving confirmed feedback that the switchgear operation has been successfully carried out. During the configuration process, the Logical Node can be instantiated with one control state machine or another. Therefore, it is the Server, the Bay Control Unit, which dictates which control model can be applied by the client.

Commands can come from the Bay (IED), the Station (Substation HMI) or the Remote-Control Centre Levels. Ordinarily a protocol other than IEC 61850 is used for the telecontrol link to the Remote-Control Centre, in which case the Substation Gateway will convert those commands to IEC 61850 control services for operations within the substation.

The question of control authority between the Bay and Station Levels and the Remote-Control Centre is of great importance for the safe operation of switchgear within the substation. It is therefore recommended that the Data Object MltLev, which would allow control authority from multiple levels simultaneously, be set to False. The status of the data object LocSta has the purpose of managing the switching authority between local and remote operation. The Originator Category (OrCat) parameter indicates from which level of control the command has come from. Optionally the orIdent parameter can be utilized to register the name of the user who issued the command.

If the client has permissions to send commands to the Switchgear, it can apply the following control services:

- Operate
- Time Activated Operate
- Cancel
- Select
- Select with Value

While these are, in the main, self-explanatory, the Time Activated Operate allows for the client to stipulate a time in the future wherein the control command would be realized. The difference between Select and Select with Value is that the later initiates a select before Operate timeout.

## 16. Process bus communication from switchgear to IED

The Sampled Value streams defined in IEC61850 -9-2 are emitted at a frequency of 4 kHz, which for a 50Hz network is 80 samples per cycle. Assuming that each Sampled Value packet contains around 150 bytes, then each stream consumes around 5 Mbps of bandwidth [17]. All other types of traffic within the SAS have a much lower bandwidth usage. but a 100 Mbps network is still very limiting for any SASs which includes a Process Bus. A 1 Gbps network should be specified for Process Bus implementations to avoid overly limiting the possible number of Sampled Value streams. The messages themselves should take no longer than 2 ms to leave the Merging Unit (or PIU). The transmission time introduced by the LAN itself, especially the ethernet switch, should if correctly configured, be measured in the hundreds of microseconds. In larger process bus implementations with lots of Sampled Value Streams, care should be taken to avoid jitter due to overly large MAC address tables within the Switch.

Many station bus installations use a ring configuration for their LAN topology and apply the Rapid Spanning Tree Protocol (RSTP) in the event of a link failure and to stop the same messages from continuing to re-circle throughout the ring formation. It works by putting some links in a standby mode such that traffic cannot flow. In the event of a link failure somewhere else in the LAN, then those links in standby mode can be re-activated to provide an alternative route. In this respect, Rapid Spanning Tree protocol is self-healing and resilient, but on its own, it is insufficient for Process Bus applications. In the event of a link failure within the Process Bus. Even the very best implementations of Rapid Spanning Tree protocol would take a minimum of 5 ms to restore communications in the event of a link failure. This is not acceptable for the sending of mission critical and time-sensitive sampled values. Therefore, the redundancy protocols specified in IEC 62439−3 are used: high-availability seamless redundancy (HSR) and parallel redundancy protocol (PRP). While their network topologies are different, the principle of operation is the same in each case, such that each message is sent twice via different routes. Therefore, in the case of normal operation, the receiving IED is prepared to take the first message which arrives and to filter out the second instance of the same message. In the case that a link failure occurs, then the receiving IED will seamlessly receive just one instance of the message, without even incurring an extra delay nor loss of information. PRP implementations involve two independent LANs operating in star formation. PRP installations have a slightly elevated cost due to the extra switches required for the additional LAN. The IEDs must be PRP prepared to be able to receive packets from both LAN as depicted in Fig. 8.12 and ignore the second instance of the same message. Due to the star formation, it is easy to include new bays or take a position out for maintenance.

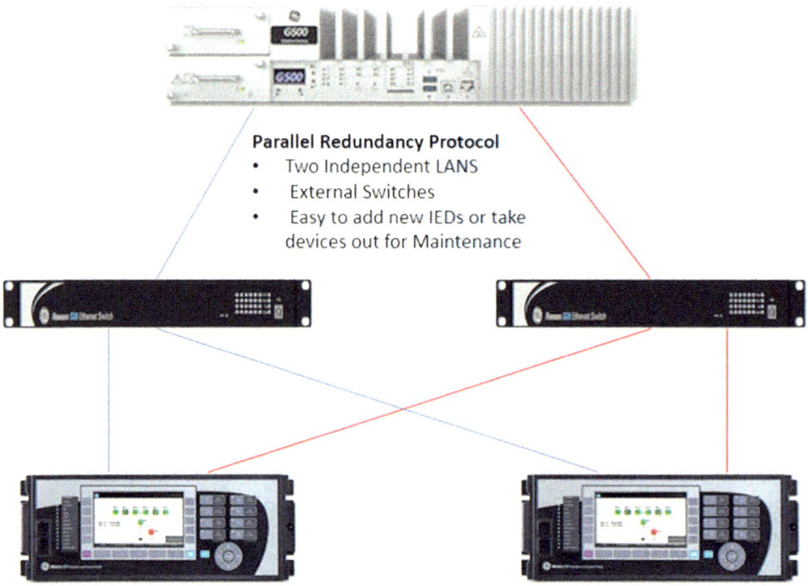


**FIGURE 8.12** Simple PRP set-up with two LANs.

HSR networks are in ring formation as shown in Fig. 8.13. There is no need for an external switch. The IEDs are looped together in a ring, wherein the IEDs themselves act like a switch. The messages are sent in opposite directions within the ring, such that should there be a link failure and one message path has been interrupted, the other instance of the same message will arrive to the receiving IED, without any extra delay or loss of information. These HSR networks have some disadvantages in comparison with PRP, there is a loss of redundancy whenever the ring is opened for maintenance, bay extension or device failure, the IEDs must have HSR switching capability and of course by duplicating each message on the same LAN, then the bandwidth

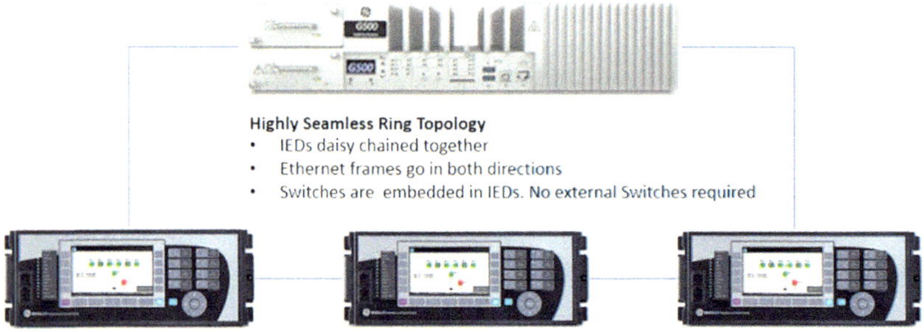


**FIGURE 8.13** Simple HSR daisy chain topology.

requirements are also increased. HSR is a very popular option for distributed busbar differential applications where each IEDs acts as a merging unit to send sampled values to a central unit.

While it is possible to have physically separate LAN for the process bus; in most cases, this extra expenditure is not justified. Having both the station and process bus traffic on the same network saves money on switches but requires that a logical separation is applied. This logical segmentation is provided by virtual LANs (VLANs), which can be used to create logically separated ethernet networks that share the same switch, fibre optic, ethernet cable and connections as with the VLANs for other traffic types. Another benefit of applying VLANs is that they support prioritization, such that Sampled Value traffic can be given the highest priority so that the Switch doesn't just empty its buffer in a First in First Out approach but will instead give priority to Sampled Value messages over GOOSE messages and Goose traffic over MMS traffic and MMS traffic over management traffic. This reduces the maximum latencies and jitter in the networking transmission time on the process bus.

It is not enough however that the Sampled Values Packets are received by the protection in 2.5 ms. It is also necessary that time synchronization errors are not greater than 1 μs. Simple Network Time Protocol (SNTP), while a valid solution for station bus only implementations, has an accuracy of 1 ms and is thus not viable for Process Bus installations. Precision Time Protocol (PTP), defined by the IEEE 1588 standard and depicted in Fig. 8.14, is the only time synchronization method which uses the existing LAN and is able to provide the required synchronization accuracy. The use of PTP within IEC 61850 substations is covered by the part IEC 61850-9-3. It operates on a hierarchy of clocks, where the grandmaster could be for example the Global

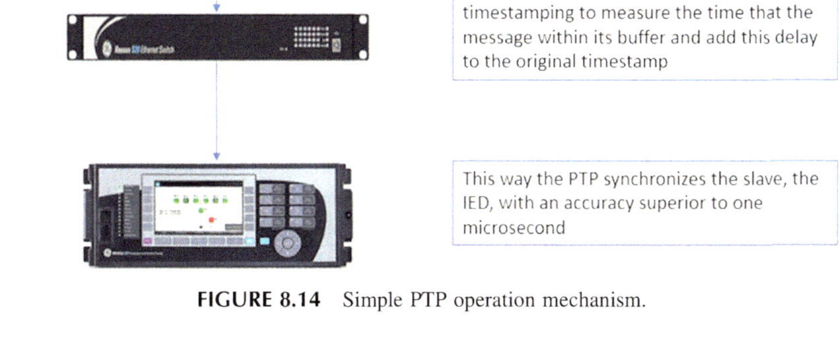


**FIGURE 8.14** Simple PTP operation mechanism.

Positioning System (GPS) receiver and the Ethernet Switch itself could be classed as Transparent Clock and the IED could be the Ordinary Clock. The Transparent Clock, in this case the Switch, measures the time the message takes between entering and leaving. Therefore, the Ordinary Clock, the slave of the PTP system, can compensate for most of the network transmission delay and have a corrected time stamp. How that correction occurs depends on whether PTPversion 1 or PTPversion 2 is being used. With PTPversion 1, the switch sends a follow-up message which states the network delay, the IED itself makes the correction using both messages. While with PTPversion 2 the switch has to correct the timestamp directly within the original message before it is received by the IED.

## 17. Switchgear maintenance using IEC 61850

To understand the asset health of switchgear, to prolong asset life and to avoid the destructive breakdown of switchgear, it is necessary to apply some level of condition monitoring. In gas-insulated switchgear, it is of utmost importance to detect any possible gas leaks, which could cause environmental problems and unsafe conditions for switching. Circuit breakers can have problems with their opening mechanism especially after a long time without being opened or after many fault breaking operations, which may create wear on the contacts used for extinguishing the arc. Partial discharges are partial electrical discharges due to insulation degradation. It is called partial because an earth fault has not yet occurred, so it is therefore an excellent indicator of insulation health and allows for intervention before a more complete insulation breakdown causing a major fault occurs. Given the importance of monitoring switchgear health for maintenance, and asset management, IEC 61850 gives us some logical nodes for modelling.

IEC 61850-7-4 specifies the logical nodes that can be used, the functionalities that can be represented within the data model. There are logical nodes for supervision and monitoring, beginning with S, logical nodes for instrument transformers and sensors beginning with T, for switchgear beginning with X and for power transformers beginning with Y.

Gas-insulated switchgear can use partial discharge supervision to detect insulation failures, insulation gas supervision (pressure and temperature) to detect leakages, and the operating mechanism supervision to supervise the DC coil which is energized during circuit breaker operations [18]. IEC 61850-7-4 provides logical nodes for all these, SPDC is for the monitoring and diagnostics for partial discharges, SIMG is for insulation gas supervision and SCBR for the supervision of the circuit breaker. For example, SCBR contains data objects that can be used to inform the open/closed status of the opening coil, or to alarm whether the coils are in an abnormal state whereas the logical node for, SIMG, has data objects for notifying the abnormal pressure, density or temperature of the gas insulating medium.

In the very possible situation that these monitoring devices are supplied by different manufacturers, interoperability, without IEC 61850 would remain elusive and integration with any diagnosis or analytical software would be laboriously difficult. This is not the case when IEC 61850 is being used, because not only are the communication protocols defined, but the logical nodes for representing the condition monitoring information are also commonly defined irrespective of manufacturer.

## 18. Future developments and roadmap of IEC 61850

IEC 61850 has already published an Edition 1, Edition 2, Edition 2.1 and Edition 2.2. It is not going to stop advancing, and in the next few years, a new edition could be expected that would bring changes to communications, data model and engineering. It is probable that communications will include Switch Configuration via the IEC 61850 SCL Engineering Process. Switches are currently configured via proprietary methods and including them within IEC 61850 will allow for a more complete top-down engineering process with the guaranteed compatibility between devices that this provides. Once more strict validations are implemented by subscribing IEDs for Sampled Value and GOOSE communications, the importance of having no configuration error will be even more pronounced. As of today, GOOSE messages are deployed for peer-to-peer communication between IEDs, and this is even the case in automatism schemes [19]. Therefore, there is no acknowledgement that the receiving Positions circuit breaker is able to act upon the command, much less any acknowledgement that the automated switching operation has been successfully completed. To resolve this issue, the IEC 61850 standard will probably introduce Manufacturing Messaging Specification Client Server communication for between IEDs also. The data model will be extended especially in the creation of new logical nodes for condition monitoring applications. Basic Application Profiles already exist as part of Edition 2.1, but going forward, we can expect that these macro function groupings of logical nodes and their intercommunications, will be specified using IEC 61850 System Configuration Language (SCL). This means that the IEC 61850 engineering process will be extended to include the initial specification phase and Users will be able to provide IED Specification Descriptions (ISD) files to IED manufacturers stipulating unambiguously in IEC 61850 language what data model they require. With regards to engineering, IEC 61850 will be extended to actually configure physical aspects of the IED such as the Digital Inputs and Outputs and to specify the relation between IEDs and a Device Management System, which can remotely and reliably carry out major maintenance tasks such as a firmware upgrade.

## 19. Other communication protocols

Although IEC 61850 standard-based communication protocol mechanism has revolutionized the substation communication architectures, design and network topologies, there are other important legacy communication protocols used in substation. Few of the widely used legacy protocols still in SASs are as follows:

- Modbus
- Distributed Network Protocol (DNP 3)
- ICCP

## 20. Modbus communication protocol

Modbus [20] which initially started as a popular protocol to be used in programmable logic controllers also went on to become one of the widely accepted standardized protocols in SCADA communication. It is interesting to note that this started as a proprietary protocol by Modicon. In SCADA system, Modbus communication is used for transmitting data between master stations and Remote Terminal Units (RTUs). It incorporates the layers 1, 2 and 7 of the standard OSI model shown in Fig. 8.15 and abides by the rules dictated by the media access control (MAC). In Modbus protocol, polling is used as the MAC protocol in which the master which could be the master terminal unit sends request for communication to the slaves in a pre-defined set of rules and the slaves handshake in similar pre-set rules. This has the advantage of reliability of sending message but disadvantage of being inherently slow. If multiple times the message is not sent to the destined slave, then the corresponding slave RTU is tagged as non-responding. However, the disadvantage is that all the communication is routed through the master which makes the procedure slow. The error checking mechanism is cyclic redundancy check (CRC) in which a check value is sent in encrypted form to the receiver end where along with the data sent the CRC bit is also verified.

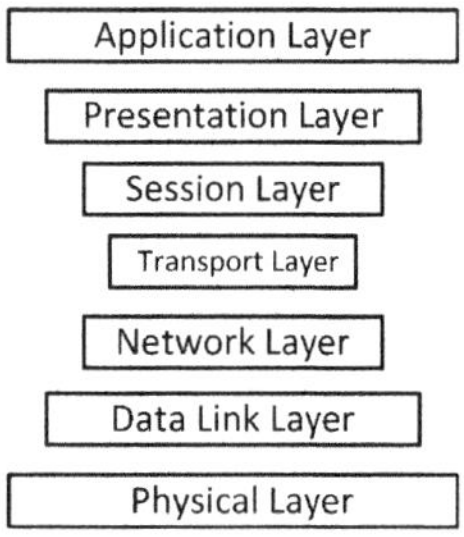


**FIGURE 8.15** Seven layers of OSI (open system interconnection) model.

## 20.1 Modbus message frame

The request sent out by the master will have a particular format or frame as shown in Fig. 8.16, which consists of 1 byte for the address of the slave which can vary from 1 to 247 and in case of a message being broadcast the address is kept as 0, the slave clarifies its own address in the response it sends. The function field which also comprises of one-byte instruction has the task that has to be carried out by the slave like reading or writing to a file or a counter. In the data field and the error check field information for completion of task by the slave and CRC check is done. In the same frame the slaves also generate responses when addressed individually; however, in case of broadcast data, no response is generated. Modbus protocol can be realized using TCP/IP and ethernet like in LAN. Modbus messages could be either query/response type or broadcast-nonresponse type. Different equipment such as HMI, PLC, Control Panel can use Modbus protocol to start a remote activity as depicted in Fig. 8.17.

# 21. DNP communication protocol

Distributed Network Protocol (DNP 3) [21] is a protocol for telemetry which is based on the Enhanced Performance Architecture (EPA). In EPA architecture as compared with the conventional seven-layer OSI model, it has three layers, namely, physical layer, data link layer and application layer. It also incorporates the pseudo-transport layer which is equivalent to the combined effect of network and transport layer of the conventional OSI model. The network layer performs the action of properly routing the flow of data from the sender side to the receiver side. The transport layer performs the tasks of the proper transfer of data between the sending and receiving end including error check.

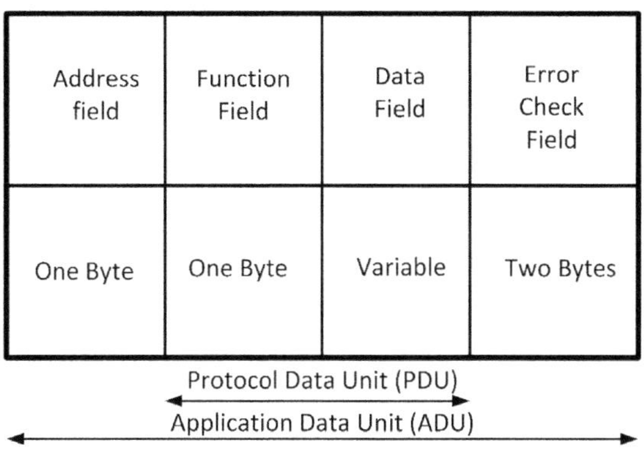


**FIGURE 8.16** Modbus data frame.

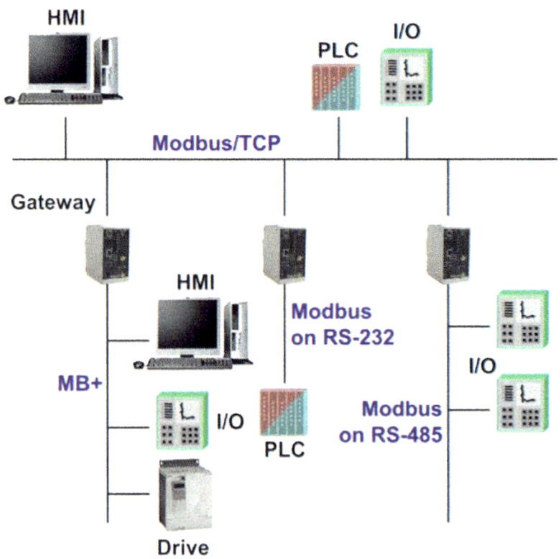


**FIGURE 8.17** Modbus transmission protocol.

## 21.1 DNP 3 structure of message frame

The starting of the frame as shown in Fig. 8.18 begins with an information regarding the data of the user, which could include an alarm signal, a variable value or any file. There is no stricture regarding the size of the data to be propagated. Application size data units (ASDUs) are the smaller sizes of the actual data for proper management. The header called as the application protocol control information (APCI) which is of length 2 or 4 bytes is appended with the ASDUs, which together give rise to the application protocol data unit (APDU). In the DNP 3 protocol, multiple master and slave topology are allowed. The purpose of DNP 3 protocol is to transmit relatively smaller sizes of data packets in a reliable manner. DNP 3 can use RS 232 connection for short end-to-end connections or RS 485 for multiport connections.

Nowadays taking into consideration the geographical areas that connect distantly located substations DNP 3 supports ethernet-based TCP/IP protocol. DNP 3 also supports peer-to-peer protocol.

# 22. Inter control centre protocol

Inter Control Centre Protocol (ICCP) [22] refers to the protocol or the specifications for communication of messages between two or more substations spread over a geographical area across a wide area network (WAN). ICCP carries the data which are real-time for supervision and control of the substations. The standard for ICCP started developing since the year 1991 by the technical committee of the IEC and Tele-Control Application and Service

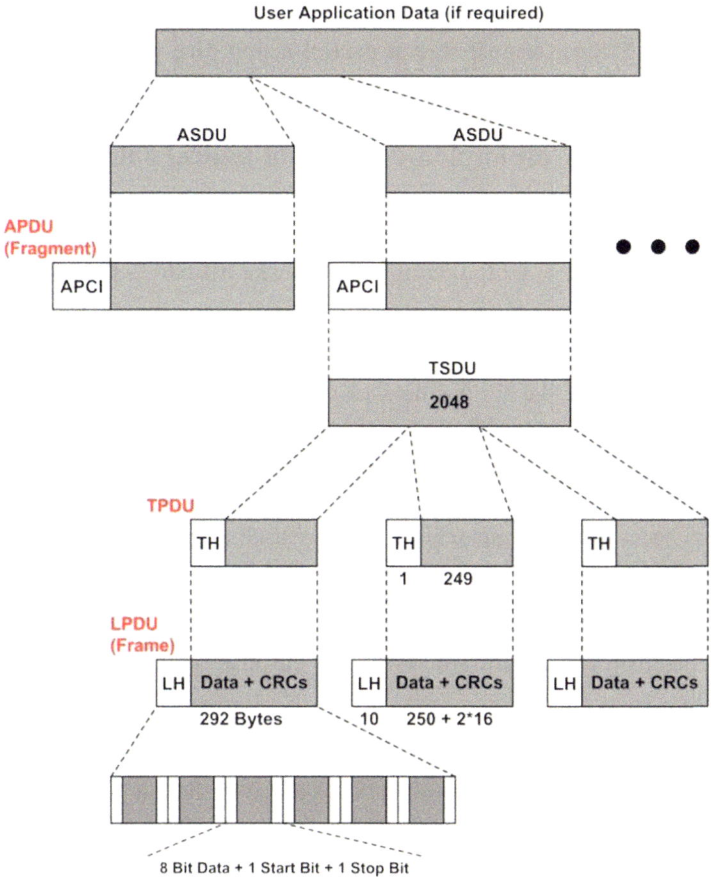


**FIGURE 8.18** Data structure of DNP 3 message frame.

Element (TASE-1) in the year 1992. Later TASE-2 released the Manufacturing Message Specifications (MMS) which is popularly known as ICCP. This is a client server protocol where a substation can pose itself as a server as well as a client depending on its role. The ICCP protocol follows the seven-layer OSI model with MMS for transfer of messages. The naming and handling of messages and decoding and interpretation of the meaning is done by Manufacturing Message Services.

The development in the communication technologies outpaces that in power system for which it becomes a necessity that the power system must be able to incorporate the changes in the Information and Communication Technology (ICT) with little effort. The traditional Modbus, DNP 3 and IEC 60870-5 methods give in to the versatility of IEC-61850, which can be extended to feeder automation as well in the distribution side. In substation, auto-reclosing operation using standby transformer peer-to-peer GOOSE communication can enable shorter control action time by replacing master-slave based Modbus communication.

## 23. Case study on implementation of typical automation scheme

In this section, a practical case study on implementation of automation scheme at typical 132/33 kV substation in India to make it smart is presented.

### 23.1 Devices used for implementation

The following are the various devices used for implementation of the automation scheme:

- **Human–machine interface (HMI):** These are the computers basically used to monitor and control the overall system in a single point. Here we can access every equipment of the substation as well as see the health of the equipment (such as transformer, circuit breaker, etc.). Two HMI units are used to maintain the reliability, as monitoring is a crucial part, and we cannott compromise with the availability of HMI at any time. So one HMI is the main system (HMI WS-M) while the other one is standby (HMI WS-S).
- **Disturbance recorder (DR):** This is also a computer in which all the logic, conditions are set, e.g. interlocking logic between isolators and circuit breakers. According to operator requirement, any new type of logic or management of new equipment can be done through this computer. This also used as a recorder to any undesirable instances occurred in the substation.
- **Gateway:** It is a router which will send the data collected from a substation to the Regional Load Dispatch Centre. It connects one substation network to another substation network. It normally works as a WAN connector.
- **GPS receiver**: This is used to provide time stamping to the data or event occurred in the field. GPS connection is provided to give a common reference of time.
- **Redundancy box (Red-Box):** Purpose of this device to combine data and time. The main role of this device is to do smooth buffering. With this smooth buffering, we can achieve proper time stamping and smooth data ow. This box also helps to convert normal ethernet to/from parallel redundancy protocol.
- **Ethernet fibre switch (EFS):** It is a junction point of the similar type of devices. Each connected device is provided with a certain address. When a data is shared, then the data frame is shared to every equipment connected to the switch as well as with other ethernet switches (depending upon the

type of networking protocol followed). Ethernet switch can facilitate full duplex data communication. Also, the device connected to it forgoes collision detection process [23].

- **Intelligent electronic devices (IEDs):** Any device incorporating one or more processors with the capability to receive or send data/control from or to an external source can be called an IED. In SAS, relays are referred to as IEDs. For each protection case, two relays are deployed: one is main relay and other one is a backup relay. Relays are numbered as per industry standard, which is a particular nomenclature that is followed. Each IED is given with a certain IP address in the LAN of the control room.
- **Bay control unit (BCU) (132 kV side):** Each substation is divided into a number of bays, based on the number of incoming and outgoing feeders are there. Each bay is provided with a BCU to control and monitor the equipment in that particular bay. It also processes the command from the operator through the HMI unit. Various logic and programs such as interlocking are set with logic gates in the disturbance recorder, which user can change as per the requirement. For future requirement, extra CIO (Contact input/Output) is also provided in BCU.
- **Auxiliary BCU**: Its purpose is to show the status of transformer health such as oil temperature indicator, winding temperature indicator, etc.
- **Bay control protection unit**: Its operation is the same as BCUs, but as this is used in 33 kV side, no separate protection unit is provided here. Protection unit is included with BCU combined giving bay control protection unit (BCPU).

## 23.2 Network model design

Overall modelling of substation networking is classified into two types as standard network and reliability network. Both types have its own application and advantages over others. So depending upon application demand and economic viability, one is preferred over other. Reliability of a network can be evaluated based on the number of point of failures, i.e. loss of communication.

- Standard network has a single failure point. This failure point can be a switch or a connecting cable. Due to these, failure communications with at least one device will interrupt. These types are cheap to set up and maintain and easiest to troubleshoot. We can use this type of network for networks, which are not much affected by time and data loss.
- On the other hand, if number of failure point is more than one, then reliable network is used. As the failure of one component will not affect the communication, it can be used for time and data critical networks. Implementation of these network types is expensive. Also, maintenance and troubleshooting is also very difficult.

In this substation, PRP [24] has been followed. Increasing dependency on power supply reliability is much more important. So continuous monitoring of the health of the network is very much important. For this, network should fulfil the following requirements:

- Successful ethernet usage in a critical network requires zero loss redundancy. This means their can not be a link or a node that is a single point of failure.
- There is also a need for zero-time recovery such that no data packets are lost.

These two important aspects of network can be achieved with PRP. It doubles every element used in the network to increase reliability. It has zero downtime during maintenance, i.e. any device can be disconnected and replaced without breaking network connectivity. Here each device is connected to the ring network through Ethernet Fibre Switch in a double star topology. In this way, each node is connected to two separate parallel LAN (ring structure). Source node always sends data to both the parallel LAN. After the data packet reaches, the destination of the duplicate data is eliminated. As all the activity happens inside the control room, PRP is not visible to the upper layer of networking and hence does not affect that in any way including the IP layer.

## 24. Automation scheme design

Metering data from the substation is taken into the control room with the help of copper cables. All the communication between the control room and field is done through copper cables, whereas inside control room communication between devices is done through fibre-optic cables. From field, signals are collected for measurement, control and protection purpose. Different copper wires are laid for different signals. Each copper cable from the field has 12 cores in it, and each core is used for a particular purpose. This way different signals are allotted for different copper wires. Now the protection signals are given directly to IED, whereas the measurement and control signal is given to BCU through the terminal block. Through BCUs and PRP network, it is communicated to the HMI via fibre-optic cables and EFS. On the occurrence of a certain event in the field, the generated soft signal will hold the data packets and send it in the PRP network. In this process, the data packets are traverse through each and every connected equipment, but it will be taken and acknowledged by only the destination device. In case the operator gives a command to any equipment of BCU, first it is processed in the set logic or programme inside the disturbance recorder and then a signal is passed to BCU through EFS. Recognition of BCU is done by the set IP address, which comes with the data frame as a soft signal in the LAN. Similarly, each element and each control signal type is provided with a particular code inside BCU, which is followed throughout the control room causing proper distinction between different signals. After BCU understands the motive of the signal, it makes logic high at the same corresponding signal in field side terminal box. This high signal will reach in the field through the copper wire, and hence, necessary action undergoes in the field.

## 25. Implementation details of SA

The CT and PT present at the field are connected with the relays in control room through copper wires. They continuously provide the relays with current and voltage values.

The relays contain two files:

- Setting files: It contains the various parameters required for fault calculation. It includes CT and PT ratios, the operating voltage, etc. In case of distance protection, the line length and conductor used (to calculate per km resistance and reactance and line angle) are included.
- Configuration files: Since the output of relays is binary, it is configured with different faults, i.e. each binary output is assigned a fault and an LED, which glows during the occurrence of that fault.

Substation automation and integration in overall can be layered into five levels [25] as shown in Fig. 8.19:

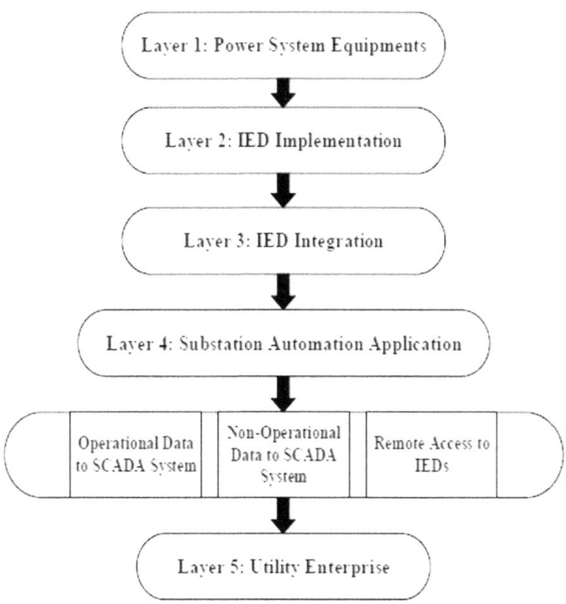


FIGURE 8.19 Levels of automation.

- The first level is the part where primary field equipment placed. The function of this level is to produce an analog signal from equipment and converts it into binary status and control signals.
- The second level is a very critical stage of automation implementation, i.e. replacement of the electromechanical relay with IEDs. As compared with electromechanical relays, these IEDs not only can execute the control command but also able to communicate data and any required information.
- The third level is dedicated to the integration of IEDs. Integration of IEDs with data acquisition, protection and control has come with lots of advantages. With IEDs, data acquisition is simplified by minimization of the number of platforms; thus resulting set-up and operational cost also reduce the control room space requirement.
- The fourth level is informing substation automation applications such as relay coordination, adaptive relaying and intelligent alarming.
- The fifth level is the utility enterprise that consists of third-party enterprise software, in which both operational and non-operational data are being analyzed. In addition to data analysis here, the analyzed data in comparison with previously stored data helps to predict the required time for maintenance instead of periodically scheduled maintenance and hence reduces the maintenance cost. The primary functional data collected from substation can be sent to the utility enterprise by three paths. The first one is via a SCADA system, the second data path is via data warehouse and the third is remote access to IEDs [26].

## 26. Conclusion

Substation automation has emerged a lot over the past decade and one of the important factors behind it is the incorporation of advanced communication protocols, architecture and design in substation operation. Communication protocols such as Modbus, DNP, etc. have been used extensively earlier for communicating the data to remote control centres, SCADA, HMI. With the emergence of IEC 61850 standard as global standard for substation automation, it revolutionized the substation communication architecture and network topologies for high-speed communications including peer-peer communication mechanisms. As the data inside the substation continue to grow with the introduction of new IEDs, PMUs, etc., communication protocols along with routing mechanism play a vital role in remote and unmanned operations of substation [27].

## References

[1] Types of substation. https://www.elprocus.com/what-is-a-substation-definition-types-of-substations/. (Accessed 25 January 2024).

[2] IEC 61850 Process Bus. http://www.pacw.org/nocache/en/issue/june_2016_issue/iec_61850_update/iec_61850_process_bus.html. (Accessed 25 January 2024).

[3] IEC 61850 Protocol Manual. https://library.e.abb.com/public/fe4d2c659c6ccad2c125783a00454ae3/1MRK511242-UEN_-_en_Communication_protocol_manualIEC_61850650_series IEC.pdf. (Accessed 25 January 2024).
[4] Smart Grid—Reinventing the electric power system. https://www.ieee-pes.org/images/files/pdf/2012-pe-smart-grid-compendium.pdf. (Accessed 25 January 2024).
[5] H. Heine, P. Guenther, F. Becker, New non-conventional instrument transformer (NCIT)—a future technology in gas insulated switchgear, in: 2016 IEEE/PESTransmission and Distribution Conference and Exposition (T&D), 3–5 May 2016.
[6] IEC 61850—Power Industry Communication Standard. https://www.automation.com/library/articles-white-papers/articles-by-bill-lydon/iec-61850-power-industry-communications-standard. (Accessed 25 June 2019).
[7] The smart switch and its supporting role. https://www.engerati.com/article/smart-switch-and-its-supporting-role-smart-grids. (Accessed 25 January 2024).
[8] Solid state transformers. https://www.pespublications.ee.ethz.ch/uploads/tx_ethpublications/spec_auckland_2016_sst_tutorial_PRESENTED.pdf. (Accessed 25 January 2024).
[9] IoT Penetration. https://ieeexplore.ieee.org/iel7/35/7452253/07452256.pdf. (Accessed 25 January 2024).
[10] M.A.K. El-Shafei, M.I. El-Hawwary, H.M. Emar, Implementation of fractional-order PID controller in an industrial distributed control system, in: 2017 14th International Multi-Conference on Systems, Signals and Devices(SSD), 28-31 March 2017.
[11] IEC 61850 Standard Overview. https://www.iec.ch/smartgrid/standards/. (Accessed 25 January 2024).
[12] T. Kostic, O. Priess, C. Frei, Understanding and using the IEC 61850: A case for meta-modelling, Art. Comp. Stand. Interf. 27 (6) (June 2005) 679–695.
[13] Modern Design of IEDs and Functions. https://w3.siemens.com/smartgrid/global/en/resource-center/Documents/Modern%20Design%20of%20IEDs%20and%20Functions-Impact%20of%20IEC%2061850.pdf. (Accessed 25 January 2024).
[14] IEC 61850 Communication Networks and Systems in Substations: An Overview for Users. https://www.gegridsolutions.com/multilin/journals/issues/spring09/iec61850.pdf. (Accessed 25 January 2024).
[15] M.C. Janssen, The visual specification method—Making specifying IEC 61850 based system easier, in: 11th IET International Conference on Developments in PowerSystems Protection (DPSP 2012), 23-26 April 2012.
[16] S.-J. Chen, Y.-H. Wang, C.-H. Lin, T.-S. Zhan, R.-F. Chang, Y.-C. Chang, Using multi-vendor IEDs for IEC 61850 interoperability and HMI-SCADA applications, in: 2012 International Symposium on Computer, Consumerand Control, 4-6 June 2012.
[17] Protection and testing considerations for IEC 61850 sampled-values based distance and line current differential schemes. Chase, S., Jessop, E., et al. 72nd Texas A&M Annual Conference for Protective Relay Engineers.
[18] Jiang, J., Using IEC 61850 for switchgear online condition monitoring. International Conference on Logistics Engineering, Management and Computer Science. (LEMCS 2015).
[19] N. Nguyen-Dinh, G.-S. Kim, H.-H. Lee, A study on GOOSE communication based on IEC 61850 using MMS ease lite, in: 2007 InternationalConference on Control, Automation and Systems, 17-20 October 2007.
[20] "Modbus Specifications. www.modbus.org/specs.php. (Accessed 25 January 2024).
[21] DNP Specifications. https://www.dnp.org/About/Overview-of-DNP3-Protocol. (Accessed 25 January 2024).

[22] ICCP Specifications. https://www.energy.gov/sites/prod/files/oeprod/DocumentsandMedia/19-Secure_ICCP_Integration.pdf. (Accessed 25 January 2024).

[23] Collision Detection. https://www.sciencedirect.com/topics/computer-science/collision-detection. (Accessed 25 January 2024).

[24] PRP protocol overview. https://ieeexplore.ieee.org/document/6699845. (Accessed 25 January 2024).

[25] J.D. McDonald, Substation automation. IED integration and availability of information, IEEE Power Energy Mag. 1 (2) (March-April 2003).

[26] A.G. Phadke, T. Bi, Phasor measurement units, WAMS, and their applications in protection and control of power systems, J. Modern PowerSyst. Clean Energy 6 (4) (July 2018) 619−629.

[27] M.S. Thomas, J.D. McDonald, Power System SCADA and Smart Grids, first ed., CRC Press, 2020.

Chapter 9

# Cybersecurity: Protection from cyber threats

**Tarlochan Sidhu[1] and Bhaveshkumar R. Bhalja[2]**
[1]*Ontario Tech University, Department of Electrical, Computer & Software Engineering, Oshawa, ON, Canada;* [2]*Indian Institute of Technology Roorkee, Department of Electrical Engineering, Roorkee, Uttarakhand, India*

## 1. Introduction

During the evolution of the power system, a remarkable transition has been observed from small, isolated networks to wide-scale interconnected ones ranging from multiple regions and countries to even across continents. The evolution also includes merging renewable energy sources with conventional sources. Though the evaluation has improved the observability and reliability of the system, it has imposed significant challenges in protection, control and optimal power management. To address the said issues, engineers have integrated communication networks with the existing power infrastructure. Though the amalgamation of communication network with power infrastructure resolves many issues, it creates new challenges in the form of cyber attacks that hinder the safe and reliable operation of power systems.

### 1.1 Evolution of communication technology in power system

Although the world witnessed a massive boon in digital and communication technology in various sectors, the power industry was relatively slow and reluctant to adapt initially. The risks and costs involved in the integration were the primary reasons for the resistance. The shift in focus and demand for a more efficient, fast and reliable system arose when, even though conventional and renewable sources were sufficient to meet the energy needs at that time, the ever-growing demand for electricity required more efficient power transmission from the existing sources. The addition of new lines is often a costly option and faces opposition from residents. Hence, communication systems were adopted and integrated with the power system to overcome the aforementioned hurdles, achieving better grid use and interoperability. Initially,

**Switchgear Design, Operation, and Maintenance Using Industry Standards**
**https://doi.org/10.1016/B978-0-443-24718-7.00009-3**

telegraphs and telephones were used in local and isolated grids. Consequently, Supervisory Control and Data Acquisition (SCADA) systems were introduced in the mid-20th century for remote control and monitoring of power transmission and distribution network. Hence, the phase of incorporating Wide-Area Monitoring, Protection and Control (WAMPAC), which includes real-time monitoring from geographically scattered networks, started.

The transition from analogue to digital in the communication system involved substitution of conventional copper wires with optical fibres. This advancement has facilitated quicker, more dependable and instantaneous monitoring of power systems, resulting in the development of the smart grid sector. Smart grids have shown superior results in terms of demand response, pricing and integration of green energy sources. On the other hand, cybersecurity becomes vital as the communication infrastructure may be exposed to malware and unauthorized users that threaten to disrupt the safe operation of the grid.

## 1.2 Cyber-physical systems

Cyber-physical systems (CPS) are an Industry 4.0 creation, which is a fusion of physical and communication systems [1]. It has helped in advancements in the field of automation, providing fast and efficient real-time monitoring and control. In addition to the power industry, CPS has served various industrial sectors such as automobile, aviation, defence, agriculture, etc. It also assists in cybersecurity for the data and control actions flowing across the communication channel.

### *1.2.1 Architecture of a cyber-physical system*

A CPS, which is an integration of a physical system with a cyber system, involves physical processors, sensors, actuators, a communication medium and a central control unit. Its architecture is depicted in Fig. 9.1. The physical system encompasses tangible processes, sensors and actuators, while the cyber system encompasses communication networks, computing resources and control centres. The physical processes are typically regarded as the system that the cyber components control.

As CPS is a complex integration of various components that work together to bridge the physical and digital worlds, an overview of these key components is presented in the following [2]:

- ***Sensors*:** Sensors are the eyes and ears of a CPS. They are responsible for collecting data from the physical environment. They can measure various parameters such as temperature, pressure, humidity, voltage, current and much more. The data collected by sensors serves as the input for the cyber system, enabling it to make decision and respond to changes in the physical world.

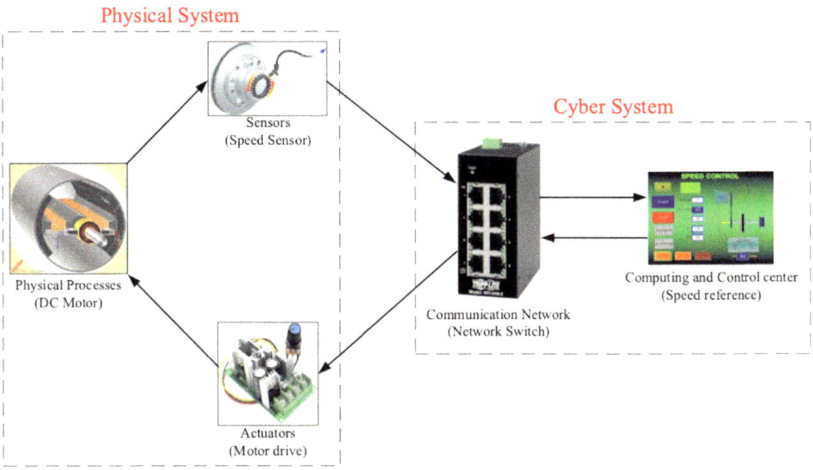


**FIGURE 9.1** Architecture of a cyber-physical system.

- ***Actuators***: Actuators comprise equipment that execute the commands as instructed by the cyber system. The action can be in the form of opening/closing the line by a circuit breaker and/or increasing/decreasing the motor speed through a controller. Actuators convert the command from the control centre into physical action in the actual system.
- ***Control and computing centre***: It is the core of the CPS and involves data centres, algorithms, analytics, visualization tools, etc. It receives data from the various sensors present in the physical system through the communication channel and analyzes the data. It decides the requirement of corrective action as per the algorithm and sends the commands back to the physical system. Based on the computational capability of the control centre, the main advantage of CPS is its ability to process data from various channels at a time with great speed and accuracy and collectively executes decisions based on the state of the entire system.
- ***Communication network***: The communication network involves both wired and wireless media. The wired communication usually consists of ethernet cables for short distances and fibre-optic cables for longer distances. Conversely, the wireless communication can be in the form of Wi-Fi, ZigBee, LPWAN, 4G, etc. It takes the data in various protocols from sensors to the control centre and returns command to the actuators.

All the aforementioned components of CPS work in tandem to respond against changes in the physical system and act upon it. It helps in automation, optimization, wide-area monitoring and better decision-making in various applications in smart grids, automobiles, healthcare, etc.

### *1.2.2 Communication infrastructure of power grid*

As power grids embrace digital transformation, operators face the challenges in terms of adaption of an enhanced communication demands and exploration of new communication avenues. Consequently, there is a noticeable shift towards integrating communication capabilities into power grids. A simplified view of the standard communication set-up within an information technology (IT)/ooperational technology (OT) network [3] is shown in Fig. 9.2.

A distinct separation exists between the office network and the process control network (PCN) within a grid operator's network. The office network resembles a standard corporate network, primarily handling tasks such as email communication and data processing. In contrast, the PCN serves as the bridge connecting the grid operator's command centres with substations and field devices. The typical protocols used in this communication are Distributed Network Protocol 3 (DNP3), MODBUS, IEEE C37.118 (Synchrophasors), IEC-61850 (SV, GOOSE), etc.

Usually, a grid operator's network has a partition between the office network and the PCN. The office network is meant for processing tasks such as email and data. On the other hand, the PCN has a programmed logic controller (PLC) that reads the control messages and sends them to the process layer. Typically, the control room has a human–machine interface (HMI), a database server (DB) for managing grid-related data and a simulation server for preemptively assessing the impacts of grid modifications. Several substations are connected to the control room, and each substation has at least one gateway, one HMI and several PLCs. As the control room may be connected to other TSO/DSO systems, it can control each system. Transferring data across the office network over to the PCN involves a specialized data exchange server that scans them for viruses before passing. However, as evidenced by the attacks in Ukraine, other communication paths such as virtual private networks (VPNs) are available. These enable contractors/suppliers to connect directly between the office network and the PCN or remote maintenance lines. However, this VPN connection may also be subjected to unwarranted access by attackers to gain confidential information.

## 1.3 Need for cybersecurity in power systems

Providing an uninterrupted and quality electric power supply has been the ultimate objective of the power industry. As a result, the system has seen rapid advancements, one of which is the adoption of the digital age. However, as seen on numerous occasions, every new technology comes with its own hardships. This is the case of the recent occurrences of various cyber attacks that cripples the smooth functioning of multiple applications in automation, protection and control. Hence, familiarizing themselves with the data, different protocols, communication channels, intelligent electronic devices (IEDs),

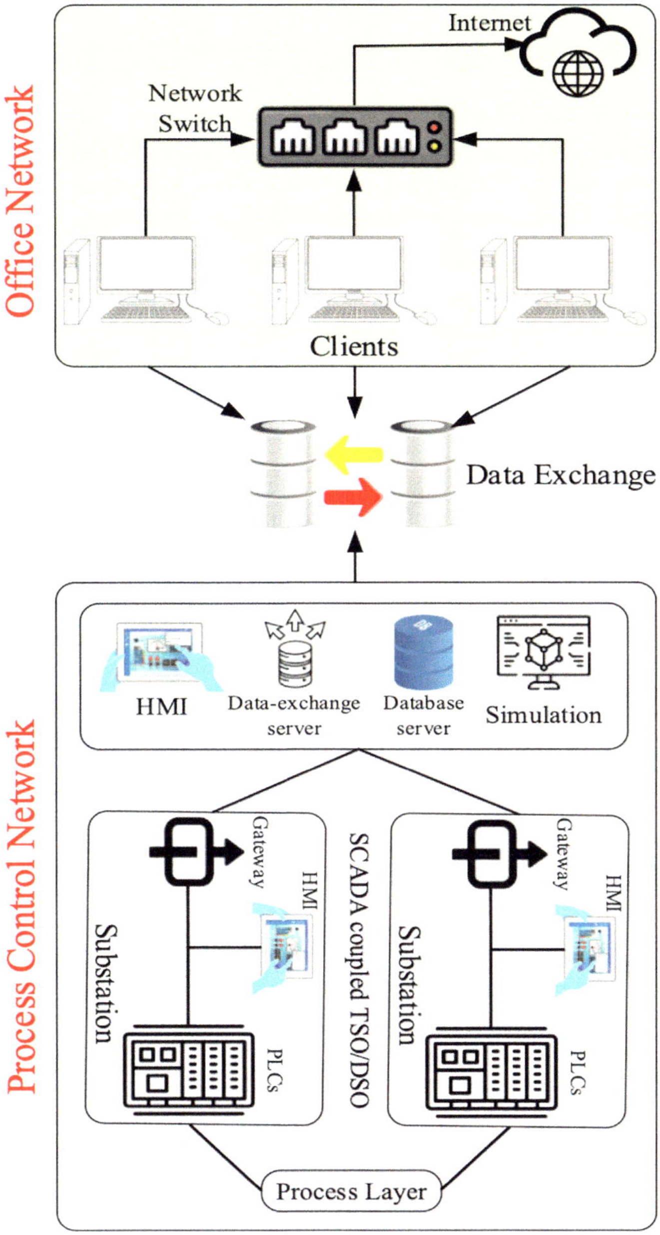


**FIGURE 9.2** A simplified view of the TSO/DSO network.

cryptography, etc., have been the need of an hour for power engineers. It makes cybersecurity a vital part of CPS's involvement in the power system. Understanding how an attacker targets a particular device or part of the network, security measures must be adaptive and resilient to detect and protect the network preemptively.

### 1.3.1 Classification of cyber attacks

Cyber attacks and their countermeasures can be classified broadly into three categories based on the intent of an attacker and the way the data flow is targeted [4]. These categories are discussed in the following:

1. **Availability-based attacks:** These types of attacks target the data flow in such a way that the data is made unavailable at the receiver's end. An attacker gets into the network and either disrupts the service or floods the network due to which the connection is overwhelmed and disconnected. Typical types of attacks in this category are denial-of-service (DoS), distributed DoS (DDoS), jammers, etc. are the typical types of attacks in this category.
2. **Integrity-based attacks:** These types of attacks target the data flow in such a way that the credibility of data is compromised at the receiver's end. An attacker allows the data to flow in the network but changes the content or injects its own data into the payload to gain an unfair advantage or deceives the receiver into functioning as intended by the attacker. Data manipulation, false data injection (FDI), sparse or replay are attacks fall under this category.
3. **Confidentiality-based attacks:** These classes of attacks target the unauthorized access, exposure or disclosure of sensitive information, compromising the confidentiality and privacy of data. They may manifest in any part of a system as attackers seek to compromise diverse system information. Tactics employed in such attacks include eavesdropping and the coordinated use of availability and integrity attack methods.

### 1.3.2 Recent occurrences of cyber attacks in CPS

In an era marked by accelerated digitalization of critical infrastructure, the energy sector has become increasingly vulnerable to cyber threats. The integration of interconnected technologies has exposed vulnerabilities that can be exploited by malicious actors, raising concerns about the security of power systems. The following instances illustrate well-documented cyber attacks on CPS, showcasing the intricate nature of these incidents and underscoring the imperative of robust cybersecurity measures to safeguard essential infrastructure:

1. **Ukraine power grid attack:** In December 2015 and again in December 2016, Ukraine experienced significant power outages. Cybersecurity researchers attributed these incidents to state-sponsored actors, pointing to the use of sophisticated malware. The attackers targeted the SCADA systems, which control and monitor the power grid. The malware used in these

attacks was named 'Black Energy' and 'Indus Troyer', and it allowed the perpetrators to remotely control and manipulate key components of the power grid [5].

2. **Stuxnet worm:** Stuxnet was a groundbreaking piece of malware discovered in 2010. While its primary target was Iran's nuclear facilities, it inadvertently highlighted the potential vulnerabilities of critical infrastructure worldwide, including power systems. Stuxnet specifically targeted industrial control systems, aiming to physically damage Iran's uranium enrichment centrifuges. Its sophistication and ability to manipulate physical processes demonstrated the potential for cyber attacks to cause real-world damage to industrial systems [6].
3. **Colonial Pipeline cyber attack:** In May 2021, the Colonial Pipeline suffered a ransomware attack by the Dark Side group [7]. The attack prompted a temporary shutdown of pipeline operations, leading to concerns about fuel supply disruptions along the East Coast. Colonial Pipeline paid a reported $4.4 million ransom in cryptocurrency to the attackers. The incident underscored the vulnerability of critical infrastructure to cyber threats and raised questions about the ethics of ransom payments, prompting increased focus on cybersecurity for essential sectors.
4. **Cybersecurity Incident on Department of Health and Human Services (HHS):** In March 2020, the Department of Health and Human Services (HHS) experienced a DDoS attack during the early stages of the COVID-19 pandemic. The attackers targeted the servers of the healthcare system to gain confidential patient health information [8].
5. **Cyber attack on Brno University Hospital:** In September 2020, a ransomware attack was carried out at Brno University Hospital targeting healthcare and patient information. The demand for massive ransom amounts was asked for the release of valuable information [9].

The aforementioned discussion suggests that the variety and the sheer volume of attacks necessitate robust cybersecurity measures, and they must be dynamic to counter any anomaly.

### 1.3.3 Implications of cyber attacks on the power system

The cyber attacks discussed before may have the capability to disrupt the power system, ranging from small-scale, like loss of a few samples of a meter, to attacks having huge impacts, such as false tripping of lines. Some of these consequences are discussed in the following:

1. **Power outages:** With the deliberate manipulation of tripping commands to a circuit breaker, an attacker can trip a line that may have cascading effects and may cause network outages.
2. **Economic effects and losses in productivity:** The interruptions caused by cyber attacks may cause discomfort and dissatisfaction to a large number of

users and also cause substantial economic losses to the utility. The resources and human resources needed to restore services also cause financial burdens.

3. **National security risks:** Power outages near military establishments and other significant government areas pose serious national security risks and may be used by rival nations to start conflicts or gain an advantage.
4. **Environmental impact:** The attacks on the control systems associated with energy sources such as nuclear energy may pose severe environmental threats. Even the remotely operated fully automated gas chamber outlets in industrial plants may cause health hazards if compromised.
5. **Loss of trust and confidence:** Considering the various consequences of cyber attacks, frequent occurrences and their adverse effects may result in socioeconomic stress and loss of trust in society.

## 2. Cyber attacks in power system

Recent cyber attack occurrences and their repercussions have become the primary concern for the engineers working in sub/power stations. We have witnessed a tremendous shift in the upgradation of existing power infrastructure with automation by incorporating more IEDs and other devices communicating with each other. The various components of the power system, starting from automated and integrated generating stations, SCADA systems, WAMS, unmanned substations and communication based smart meters installed at households, can be considered as attack surfaces for an intruder. Though developed and developing countries are investing heavily in smart grids, an equivalent amount of attention is also required towards the development of countermeasures to protect smart grids from any type of cyber attack. Various types of attacks are discussed in the following.

### 2.1 Availability-based attack

This attack class attempts to block data transmission from server to client by various means. Either disrupting the traffic or flooding the network, which results in server congestion, may cause delayed or non-delivery of important information to the client, which may be critical in taking subsequent important activity. DoS and DDoS are the most common attacks that fall under this category. Sometimes, these attacks are often referred as DoS attacks.

1. DoS attack:
   DoS attacks are the most frequent and common type of cyber attacks. This is because the attacker needs to get into the network through malware or breaching outdated firewalls and can cause disruptions without the need for knowledge of the network, protocols and the type of devices involved. The goal is to make the service unavailable to the end users, which may lead to

short-term or, in some cases, long-term disruptions [10,11]. The resources needed to detect the attack and perform corrective actions to resume healthy traffic flow may increase the financial burden on the service providers.

In recent years, the evolution of cyber threats in power systems also includes a powerful variant of DoS attacks, known as **distributed denial-of-service (DDoS)**. This requires a team of attackers that target a relatively more extensive network to isolate it from the grid completely, often referred to as BotNet. The group of attackers target different locations and control systems simultaneously by blocking a device or corresponding data traffic of each area to create a cascading effect, and collectively, a large number of users get affected. The coordinated and scattered nature of attacks makes them difficult to prevent or mitigate and may have wide-scale consequences and even national security threats [12,13].

2. Jamming attack:

   Similar to DoS attacks, jamming attacks are achieved by flooding the network with junk data so that the actual data gets delayed or the connection is lost [14]. In view of this, jamming attacks are often referred to as subsets of DoS attacks. Critical communication infrastructures such as substations and control centres are often jammed by repeatedly sending SYN (synchronize) requests. Moreover, wireless modes of communication, which are primarily used at the distribution side of the power system, are jammed by matching the frequency bands. Hence, interconnected power system operations get affected and require a restart of connection in most cases, which in turn may induce outages for a prolonged duration.

## 2.2 Integrity-based attack

This type of attack aims to compromise the integrity of the data flowing from the server to the client by altering or adding to the existing traffic flow. The reason behind the integrity of the data in question is that the data sent by a server is not the actual data received by the client. Tampering with the sensor's data, control signals or trip commands may severely affect the smooth functioning of critical devices in the power system. These attacks are more sophisticated in nature as the attacker requires in-depth knowledge of the nature of data and the network. As a result, prevention of these attacks is quite crucial and challenging.

1. FDI attack:

   FDI attacks are the type of attacks in which intruders gain access to the system and inject their own data into the system [15]. If the injected data are random in nature, an IED detects it as a conflict as compared with normal data. If the IED is unable to resolve two sets of conflicted data, it gets blocked, and hence, a crucial component of a substation is inaccessible. In this case, FDI attack leads to a DoS attack. On the other hand, if the attacker is intelligent enough and the data is equivalent to healthy data, the operators or other control systems that use these data may get tricked into making wrong decisions.

A subset of FDI attacks is the data manipulation attacks (DMAs). Here, the attacker gains access to a communication network and changes some or the entire content of the packet flow, which in turn may result in unexpected functioning of critical components of the network [16].

2. Man-in-the-middle (MiTM) attack:
   In this attack, an attacker portrayed as an authorized user gets into a network and gains access to various operations and visualization tools. Sniffing and spoofing are the main objectives of this attack [17]. Sniffing is the process in which, after gaining access, the attacker observes and analyzes the data and may use it to gain an unfair advantage in the energy market. The next step is spoofing, in which the attacker, upon analysis, decides to manipulate the data again to gain an advantage.
3. Sparse attack:
   As the name suggests, sparse attacks are achieved by inserting zeros or null values into the healthy data. For example, adding zeros to meter readings that have a higher importance in the weight matrix of the state estimation algorithm due to which it may result in inaccurate results that affect system operators' system monitoring and fault identification [18].
4. Replay attack:
   Replay or playback attacks involve saving a period of data and replaying it over and over in place of the healthy data. For instance, an attacker understands a healthy scenario and saves the data for future use. Then, the attacker decides to block the healthy data and sends the saved data repeatedly, which is similar to FDI attack. In this case, the destination device may miss even fault and other abnormal scenarios. If it is not addressed at the right time, in some cases, repeated execution of commands is observed, which may harm the device temporarily or permanently [19].
5. Manipulation of smart-meter reading:
   These attacks are the sophisticated version of the meter tampering of the early days. The attacker, in this case, alters the meter readings to reduce the monthly consumption bill. This results in a loss of revenue to utilities [20].

## 2.3 Confidentiality-based attacks

As the name suggests, this class of attacks requires unauthorized access and theft to sensitive information of crucial components of the power system. These types of attacks could be a combination of DoS and integrity-based attacks to attain system knowledge and use them to create more damage later.

1. Eavesdropping attacks:
   In this attack, network spoofing is observed by intercepting data traffic to eavesdrop and attain confidential information [21]. The data may be used to replicate a network or fault scenarios and used later to simulate an attack disguised as a fault.
2. Information leaks from network vulnerabilities:
   A lot of vulnerabilities exist in any system. Outdated and ineffective firewalls and bad practices in user authentication, such as unchanged and weak passwords, VPNs, etc., are a few of them. An attacker uses these vulnerabilities to replicate a network using data such as topology, network parameters and location of critical protection and monitoring devices and use it to simulate an attack in that network.

## 3. Cybersecurity measures

Shielding the confidential data flowing across a network is paramount to keep the grid safe and reliable. Conventional methods include rules (signatures) based anomaly detection, firewalls and set algorithms. These algorithms act upon pre-defined rules and some fixed patterns, deviating from which generate alarms. While these methods can serve as the first line of defence, they may fail in dynamic and diverse attacks. On the other hand, artificial intelligence (AI) and machine learning (ML)/deep learning (DL) methods learn from various attack patterns over time, and they excel in dynamic anomaly and intrusion detection against various emerging attacks.

### 3.1 Conventional methods

The traditional cybersecurity measures depend on the firewalls that check the network traffic and filter out unwanted players accessing the network. Even the anti-malware detection comes in this category. They rely on pre-defined rules and signatures and detect threats on some fixed patterns. By imposing access restrictions, a firewall mitigates the potential for an intrusion. Blocking underutilized and dangerous protocols and services reduces the path for attackers to penetrate a system. Another benefit of the blocking is that it might make system vulnerabilities less obvious to attackers.

1. Bad data detection in state estimation
   An essential objective of state estimation (SE) in power system operations is to recognize meters giving noisy or inaccurate and missing measurement readings. It is achieved by comparing the meter values with the states calculated by SE iteratively. Hence, less weightage is given to those meters in the weight matrix of the cost function. The SE also uses bad data detection to filter out missing or erroneous readings before passing them into SE. This prevents SE from marking a particular meter as faulty or

ineffective due to a few bad data. Bad data are detected by computing the residual, which is the difference between observed and estimated measurements, normalizing it, and comparing with a statistical threshold [22]. The measurements violating the above comparison are marked and filtered out before SE. However, a bad data detector is incapable of knowing whether the missing or erroneous values are due to a faulty meter or a cyber attack. Hence, bad data detection can act as a first line of defence against attacks.

2. Metrics for network security
   When considering an IED from the viewpoint of a potential attacker, the significance of various IED categories varies based on the attacker's objectives [23]. For example, a person intending to take down a certain section of the power system may prioritize manipulating a relay, whereas someone seeking sensitive information may target a data logging device. Additionally, diverse units will possess distinct security levels corresponding to their importance. As a result, a comprehensive framework is presented in Ref. [23] to obtain a metric to identify a particular IED's vulnerability to a certain attack as described in the following:
   1. The initial stage involves identification of potential threats to various IED categories. This is achieved by categorizing IEDs based on their designated functions and identifying potential attack scenarios, encompassing both physical and logical aspects.
   2. After pinpointing the threats to an IED, the next step involves assessing whether the device incorporates suitable security countermeasures or not. It is possible to remove a threat from the list and render it ineffective if there are enough countermeasures.
   3. A threat can go through adjustments based on its relative susceptibility. This parameter determines the probability of the attack based on the attacker's location.
   4. In the successive stage, a mathematical equation is developed to measure the level of security for both an IED and the IEC-61850 network. Hence, a comprehensive score for the entire network having multiple IEDs is obtained.
   5. Finally, the score is compared with a compliance threshold to determine the level of security for a particular network.
3. Intrusion detection system for automated substations
   Intrusion detection (ID) involves identifying a malicious intruder either during an attempt to breach or after successfully entering a protected system. Contemporary intrusion detection systems (IDS) can be categorized based on their detection methods [24]:
   1. Based on rules (signatures).
   2. Based on anomaly detection.

   Rule-based ID operates under the assumption that some types of attacks display clear, identifiable indicators, such as when a user tries to access a file without authorization or uses a brute-force password break. On the other hand, anomaly detection primarily identifies masquerades, wherein the intruder disguises themselves as a legitimate user. Presently, anomaly detection-based IDS employ statistical methods or machine learning for classification.

IDS can be further categorized based on their organization into host-based (standalone) or distributed systems. A host-based IDS analyzes the system calls of its own operating system and responds to suspicious calls. On the other hand, a distributed IDS functions across an entire network, adopting either a centralized or decentralized approach. Hybrid systems amalgamate the advantages and disadvantages of both types of systems.

## 3.2 Machine learning and deep learning-based methods

Moving away from the conventional methods that rely on set rules and patterns, a boon in ML and DL algorithms in various sectors has enabled power system cybersecurity engineers to apply these methods in CPS-enabled power transmission and distribution systems. The ability of the ML and DL algorithms to interact with a considerable volume of data and learn from these historical sets to analyse different patterns involved in the data makes them quite suitable to counter various cyber attacks due to their dynamic nature. Some methods need explicit input of features of a particular data set, and some have the inbuilt capability to extract their features. Hence, it is possible by these algorithms to identify patterns that went unnoticed by conventional methods.

1. Detection of FDI and FCI attacks using multilabel ML classifier algorithms
   In the recent shift towards wide-area monitoring, protection and control applications, a crucial communication occurs between a phasor measurement unit (PMU) and a phasor data concentrator (PDC). A PMU might be placed in a substation of a power transmission system, and a PDC might be located in the control centre. In this communication, PMU acts as a server and PDC as a client. A corresponding emulation network is designed to replicate the PMU-PDC communication architecture involving IEEE C37.118 (Synchrophasor) protocol that enables real-time monitoring of phasor data. The network is simulated with both healthy and attack scenarios, and the data is aggregated to be used by ML algorithms that are trained to analyse the distinct patterns [25].
   Various ML algorithms such as *K-nearest neighbour (k-NN)*, *decision tree (DT)* and *naïve Bayes (NB)* undergo training using a consistent approach of splitting data into training and test sets. This compiled dataset is shuffled to evenly distribute labels from individual datasets in the testing and training data. Following this, the ML classifiers are trained using the training data.

Each ML algorithm is then tasked with predicting the label, whether it indicates a type of attack or signifies normal traffic, for the test data as well. Finally, the predicted labels are compared side by side with the actual labels of the test data to calculate metrics such as the confusion matrices having false positives, true positives, false negatives, and true negatives, precision, recall, and unweighted F1-score.

2. Study of DoS-impacted PMU data using random forest classifier
   Synchrophasors are employed in ensuring the reliability of the power grid by determining the system's status, whether it is in a normal or stressed state, enabling the relaying protection logic to dynamically adjust the dependability-security bias. In these kinds of applications, ML algorithms that have been trained on massive amounts of wide-area PMU data are vital for determining the system's state. Every measurement from the affected PMU connection in the output PDC data stream becomes unavailable when a DoS attack is launched on any PMU-PDC connection. The study investigates the influence of missing PMU data on the functionality of ML algorithms, employing a well-known random forest (RF) classifier [26].

To evaluate the effects of the DoS attack, a novel testing dataset is created by implementing a reset DoS attack on the connection between PMU and PDC along with various other events such as load changes, line faults, line tripping and so on. A DoS attack reduces the number of features that the RF classifier encounters. The RF model is renowned for its adaptability to absent data, which is achieved through the utilization of an internal mechanism that selects subsets of features utilized in each data sample at random. However, there is a decrease in the accuracy of the tests, which goes from 100% to 73.2232% with the presence of DoS attack. This indicates that DoS attacks can potentially distort the RF algorithm's classification capability, which is otherwise well suited for handling such scenarios. Consequently, there is an adverse impact on the adaptive relaying application. Enhancing the testing accuracies might be achievable by training the RF model with the missing data. Nevertheless, the impracticality and limited scalability originate from the extensive variety of possible DoS attacks and the ambiguity around the specific PMU that may be targeted. The accuracy of the RF classifier can be enhanced using one or a combination of methods such as data augmentation, feature engineering, ensemble methods, regularization techniques, transfer learning, etc.

## 4. Case study of DOS-impacted synchrophasor data and its countermeasures

Fig. 9.3 shows a typical case of an attacker penetrating a CPS commonly observed in a power transmission system. Here, multiple IEDs placed at various locations in the network are communicating with a centrally located data concentrator through a network switch. The data can be voltage and/or current phasors using IEEE C37.118 (Synchrophasor) protocol, sampled values or substation events/commands as per IEC 61850 (SV, GOOSE).

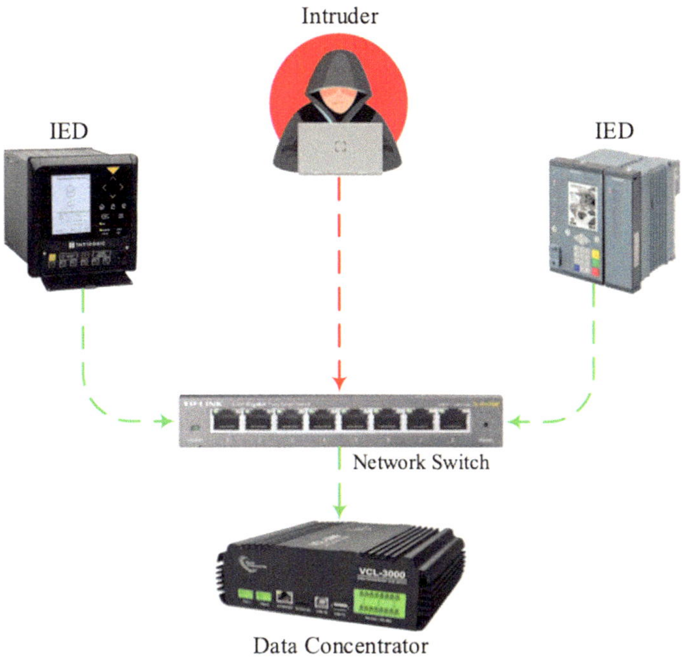


**FIGURE 9.3** A typical communication network in power system along with an intruder.

The data collected by the concentrator can be utilized for analysis, monitoring, protection and control applications. Here, through the network switch, an attacker tries to gain access to the network. This can be achieved through insecure VPN connections, malware through USBs or firewall breach. The attacker targets a computer connected to the same switch where the rest of the IEDs are connected. Once the access is established, the attacker can put a watch on the network traffic to understand the system using a packet analyser such as Wireshark, tshark, Snort, etc. A sample Wireshark capture of Synchrophasor packet flow is shown in Fig. 9.4.

Once the network and its underlying protocols are analysed, the intruder can launch a DoS attack in which the packets are dropped at the network switch. As a result, the algorithms that are using the data collected at the concentrator may get affected and give inaccurate results. The Wireshark capture during a DoS attack in the middle of a Synchrophasor data flow is shown in Fig. 9.5.

To develop countermeasures against DoS attack, a bulk of data, both in healthy and attack scenarios, can be generated. The Wireshark captures can be used to store the data, and ML can use these data for training various classifiers as mentioned in previous sections. These classifiers can be used for detection of cyber attacks. The methods based on DL can be used for mitigation of attacks so that the lost data can be reconstructed and sent to the concentrator.

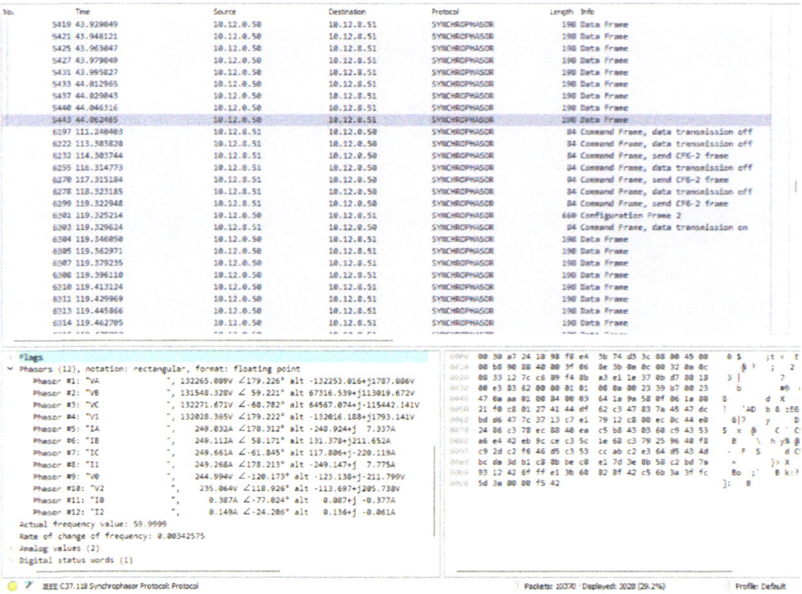


**FIGURE 9.4** A sample Wireshark capture of Synchrophasor data.

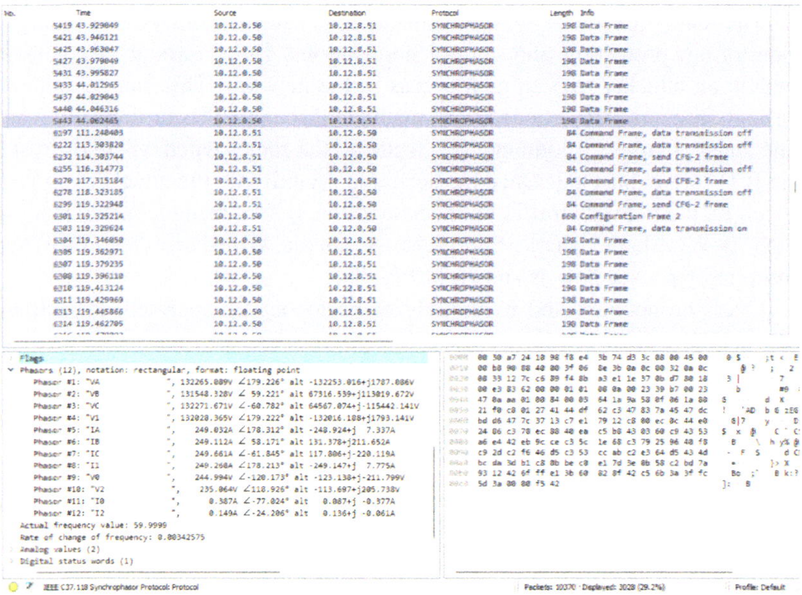


**FIGURE 9.5** A Wireshark capture of Synchrophasor data during a DoS attack.

## 5. Summary

In this chapter, a comprehensive study on the integration of communication technology with the existing power infrastructure is explored. Initially, the cyberphysical systems are introduced, which acts as the bridge between power systems and digital technology. Due to implementation of CPS, the efficiency, reliability, and flexibility of network-controlled system operation can be increased. However, it has led to new security challenges in the form of cyber attacks. Then, different classes of cyber attacks, their impacts and the recent occurrences of them across various sectors were examined. Moving forward, each attack is distinguished based on the intent of an attacker and the way it is affecting the data flow within a network. Eventually, various countermeasures, developed to address the said cyber threats, were investigated. Then, the conventional methods employing various algorithms, such as bad data detection in SE as well as the traditional rule or signature-based analysis, were studied. Though the aforementioned methods work well as a first line of defence that can filter out bad data or traffic violating certain pre-defined rules, the security measures need upgradation due to intelligence of an attacker and sophisticated nature of attacks. as well. Pertaining to this, various ML and DL methods are examined that can not only detect type of attack but also mitigate it so that the damage can be limited. Various attack prevention strategies can be explored, which can prohibit the occurrence of an attack in the first place.

## References

[1] Y. Lu, Cyber physical system (cps)-based industry 4.0: a survey, J. Indust. Integr. Manag. 2 (03) (Nov. 2017) 1750014.

[2] K.C. Sou, J. Lu, Relaxed connected dominating set problem for power system cyber–physical security, IEEE Transac. Control Network Sys. 9 (4) (Dec. 2022) 1780–1792.

[3] T. Krause, R. Ernst, B. Klaer, I. Hacker, M. Henze, Cybersecurity in power grids: challenges and opportunities, Sensors 21 (18) (Sep. 2021) 6225.

[4] A. Humayed, J. Lin, F. Li, B. Luo, Cyber-physical systems security—a survey, IEEE Internet Things J. 4 (6) (Dec. 2017) 1802–1831.

[5] N. Saxena, L. Xiong, V. Chukwuka, S. Grijalva, Impact evaluation of malicious control commands in cyber-physical smart grids, IEEE Transac. Sustain. Comput. 6 (2) (Apr.-Jun. 2021) 208–220.

[6] S. Al-Rabiaah, The "Stuxnet" virus of 2010 as an example of A "APT" and its "recent" variances, in: 21st Saudi Computer Society National Computer Conference (NCC), Riyadh, Saudi Arabia, Apr. 2018, pp. 1–5.

[7] J. Beerman, D. Berent, Z. Falter, S. Bhunia, A review of colonial pipeline ransomware attack, in: 2023 IEEE/ACM 23rd International Symposium on Cluster, Cloud and Internet Computing Workshops (CCGridW), Bangalore, India, May 2023, pp. 8–15.

[8] A.O. Ugwu, X. Gao, J.O. Ugwu, V. Chang, Ethical implications of AI in healthcare data: a case study using healthcare data breaches from the us department of health and human services breach portal between 2009-2021, in: International Conference on Industrial IoT, Big Data and Supply Chain (IIoTBDSC), Beijing, China, Sep. 2022, pp. 343–349.

[9] J. Stephens, Serious Cyber-attack Targets Brno University Hospital, Brno Daily, Mar. 2020.

[10] A.-Y. Lu, G.-H. Yang, Stability analysis for cyber-physical systems under denial-of-service attacks, IEEE Trans. Cybern. 51 (11) (Nov. 2021) 5304–5313.

[11] K.-D. Lu, Z.-G. Wu, Resilient event-triggered load frequency control for cyber-physical power systems under DoS attacks, IEEE Trans. Power Syst. 38 (6) (Nov. 2023) 5302–5313.

[12] W. Xu, J. Weng, B. Lou, X. Huang, H. Huang, J. Wu, D. Zhou, Vulnerability assessment of distributed load shedding algorithm for active distribution power system under denial of service attack, CSEE J. Power Energy Sys. 9 (6) (Nov. 2023) 2066–2075.

[13] M.A. Aladaileh, M. Anbar, I.H. Hasbullah, Y.-W. Chong, Y.K. Sanjalawe, Detection techniques of distributed denial of service attacks on software-defined networking controller–A review, IEEE Access 8 (Aug. 2020) 143985–143995.

[14] Y. Guan, X. Ge, Distributed attack detection and secure estimation of networked cyber-physical systems against false data injection attacks and jamming attacks, IEEE Trans. Signal Inform. Process. Networks 4 (1) (Mar. 2018) 48–59.

[15] G. Liang, J. Zhao, F. Luo, S.R. Weller, Z.Y. Dong, A review of false data injection attacks against modern power systems, IEEE Trans. Smart Grid 8 (4) (Jul. 2017) 1630–1638.

[16] X. Wang, D. Shi, J. Wang, Z. Yu, Z. Wang, Online identification and data recovery for PMU data manipulation attack, IEEE Trans. Smart Grid 10 (6) (Nov. 2019) 5889–5898.

[17] G. Tertytchny, H. Karbouj, L. Hadjidemetriou, C. Charalambous, M.K. Michael, M. Sazos, M. Maniatakos, Demonstration of Man in the Middle Attack on a Commercial Photovoltaic Inverter Providing Ancillary Services, IEEE CyberPELS (CyberPELS), Miami, FL, USA, Oct. 2020, pp. 1–7.

[18] M. Ozay, I. Esnaola, F.T. Yarman Vural, S.R. Kulkarni, H. Vincent Poor, Distributed models for sparse attack construction and state vector estimation in the smart grid, in: IEEE Third International Conference on Smart Grid Communications (SmartGridComm), Tainan, Taiwan, Nov. 2012, pp. 306–311.

[19] H. Guo, J. Sun, Z.-H. Pang, Analysis of replay attacks with countermeasure for state estimation of cyber-physical systems, IEEE Trans. Circuit Syst. II Express Briefs 71 (1) (January 2024) 206–210.

[20] A.P. Kamatagi, R.B. Umadi, V. Sujith, Development of energy meter monitoring system (EMMS) for data acquisition and tampering detection using IoT, in: IEEE International Conference on Electronics, Computing and Communication Technologies (CONECCT), Bangalore, India, Jul. 2020, pp. 1–6.

[21] W. Yang, Z. Zheng, G. Chen, Y. Tang, X. Wang, Security analysis of a distributed networked system under eavesdropping attacks, IEEE Trans. Circuit Sys. II: Express Briefs 67 (7) (July 2020) 1254–1258.

[22] C. Liu, W. He, R. Deng, Y.-C. Tian, W. Du, False-data-injection-enabled network parameter modifications in power systems: attack and detection, IEEE Trans. Ind. Inf. 19 (1) (Jan. 2023) 177–188.

[23] U. Premaratne, J. Samarabandu, T. Sidhu, R. Beresh, J.-C. Tan, Security analysis and auditing of iec61850-based automated substations, IEEE Trans. Power Deliv. 25 (4) (Oct. 2010) 2346–2355.

[24] U. Premaratne, J. Samarabandu, T.S. Sidhu, R. Beresh, J.-C. Tan, An intrusion detection system for IEC61850 automated substations, IEEE Trans. Power Deliv. 25 (4) (Oct. 2010) 2376–2383.

[25] K. Knesek, P. Wlazlo, H. Huang, A. Sahu, A. Goulart, K. Davis, Detecting attacks on synchrophasor protocol using machine learning algorithms, in: IEEE International

Conference on Communications, Control, and Computing Technologies for Smart Grids (SmartGridComm), Aachen, Germany, Oct. 2021, pp. 102–107.

[26] A. Chawla, A. Singh, P. Agrawal, B.K. Panigrahi, B.R. Bhalja, K. Paul, Denial-of-Service attacks pre-emptive and detection framework for synchrophasor based wide area protection applications, IEEE Syst. J. 16 (1) (March 2022) 1570–1581.

Part III

# Operation and maintenance

Chapter 10

# Standard operation and maintenance of circuit breakers

Rudra Prakash Maheshwari
*Indian Institute of Technology Roorkee, Roorkee, Uttarakhand, India*

## 1. Introduction

Circuit breakers (CBs) are mechanical devices used to isolate a specific electrical section. They receive command from actuating/master trip relay and respond accordingly. The signal from master trip relay is given to actuating coil of the CB. Generally, this is 110V signal and is used to operate a motor which is responsible to move away the contacts of CB for isolating the supply. Another feature of CB is to provide quick quenching of arc caused by separation of high current carrying elements. Based on the magnitude of current, different methods of arc quenching are utilized. This current constitutes load current along with the component caused by line capacitance and cable inductance. In interconnected system, CBs are also used to regulate power flow from generation to load [1−6].

### 1.1 Historical evolution

Prior to 1970, depending on the medium of insulation, air and oil were the only choice for arc quenching in the HVCB. Air blast CB, minimum oil CB, oil CB for voltage levels of 400, 132 and 33 kV, respectively, were available in the market. Later on, due to advancement of vacuum creating capabilities, gradually vacuum circuit breakers also found place. However, in the meantime, SF6 circuit breakers came up and find application in power applications. The sole objective of any type of CB is to minimize the amount of power dissipated due to arcing either by quickly eliminating the arc or by increasing the length of arc to increase arc resistance. Table 10.1 gives an idea of voltage levels and type of CBs used.

**Switchgear Design, Operation, and Maintenance Using Industry Standards**
**https://doi.org/10.1016/B978-0-443-24718-7.00010-X**

**TABLE 10.1 Different types of CB used at various voltage levels [3].**

| Voltage level | Category of CB | Options available |
|---|---|---|
| <1000 V | Low voltage | Miniature circuit breaker (MCB) |
| | | Moulded case circuit breaker (MCCB) |
| | | Air circuit breaker (ACB) |
| 3.3–36 kV | Medium voltage | Vacuum circuit breaker (VCB) |
| | | Vacuum contactor (VC) |
| 36–132 kV | High voltage | Minimum oil circuit breakers (MOCBs) |
| | | Sulphur hexafluoride ($SF_6$) |
| 220 kV and above | High voltage | Sulphur hexafluoride ($SF_6$) |
| | | Non—$SF_6$braker[a] |

[a]*Note—Non $SF_6$ CBs are under research and commissioned at very few places [3].*

Arc quenching technologies used in CBs have revolutionized the protection of electrical network. In modern power system, auto reclosing is employed after the occurrence of fault to enhance the power system reliability. This allows the poles of CBs to be closed one by one, leading to significant changes in the protection system.

## 1.2 Circuit breaker phenomenon

### *1.2.1 Short-circuit breaking current of circuit breaker*

The maximum current which a CB can withstand under short-circuit condition is cleared by opening breaker contacts. This current in CB will cause thermal as well as mechanical stress in various parts of CB. Therefore, the contact area and cross section of current carrying elements are designed to withstand the stresses. Despite all precautions being taken, sometimes damage may take place to contacts or insulation of CB.

According to Joule's law of heating, the rise in temperature is proportional to the square of the current flowing through it and contact resistance. Thus, during the short circuit, this rise will be quite high. Further, this current continues to flow as long as the contacts of circuit breaker open to isolate the faulty portion. Thus, thermal stress will be proportional to the time duration for which the short circuit persisted. Therefore, the operating time of circuit breaker decides the breaking capacity of the circuit breaker.

Different metals start showing deterioration in their mechanical properties at different temperatures. Aluminium becomes soft at 160°C.

Thus, this temperature can be taken as limiting value of temperature rise of circuit breaker contacts during short circuit. Therefore, short-circuit breaking capacity or short-circuit breaking current of CB is maximum current that can flow through the CB from the time of inception of short circuit to the time of isolating the short circuit without any permanent damage to the CB.

Short-circuit condition not only causes electrical stresses but mechanical stresses as well; thus, to determine short-circuit capacity, it also requires to be taken into consideration. As a matter of practice, the short-circuit breaking capacity is expressed as RMS value of current. Further, the selection of circuit breaker also involves determining the fault current level at that point of CB.

### 1.2.2 Rated short-circuit making capacity

Unlike short-circuit breaking capacity expressed in terms of RMS value of current, the short-circuit making current is expressed in terms of peak value of current. When fault occurs, the value of fault current may rise up to twice the symmetrical fault current level. Moreover, switching on a CB before fault clearing the fault makes the faulty section of the system connected to the source. The current is the maximum in circuit during the first cycle after the breaker is closed. This is about twice of the amplitude of symmetrical fault current. Thus, the breaker's contacts are required to withstand this current during first cycle of reclosing the fault. Hence, the breaker is required to withstand short-circuit making capacity. Due to the convention that making current is expressed in peak value, the making current is always more than (2.5 times) rated short-circuit breaking current.

### 1.2.3 Duty cycle of circuit breaker or rated operating sequence

For isolating the faulted part of CB, it is required to mechanically remove the contacts; hence, the operating sequence of a CB is given by (10.1).

$$\mathrm{O} - \mathrm{t} - \mathrm{CO} - \mathrm{t'} - \mathrm{CO} \tag{10.1}$$

where 'O' indicates operation to open the contacts of CB, and 'CO' indicates closing operation of contact immediately after an opening operation without any intentional time delay. As each operation of circuit breaker results in generation of heat, if many operations are performed in quick succession, it may damage it. Therefore, a recovery time 't' is given between two operations which is by default is 0.3 s until and unless specified otherwise.

### 1.2.4 Rated short time current

In the event of temporary fault, CB is allowed to carry higher current than rated value without any failure/damage for a short duration to ensure reliability. This value of current is defined as rated short time current. As previously explained, there is always an intentional delay between fault occurrence and actual operation of CB. This delay is used to establish proper relay operation and coordination. In the event of any CB maloperation, the fault will be cleared by next CB in hierarchy. Thus, after fault occurrence, CB should be able to withstand short-circuit current for a short duration. This short-circuit current will have two major effects on CB:

1. Due to high current, thermal stress will be high, leading to increased stress of conducting and insulating parts of CB.
2. Another effect will be high mechanical stress in different parts of CB.

CBs are designed to withstand such stresses for short period of time, decided by coordination time of protection system. But short-circuit current should not be allowed for longer periods. Thus, it is expected that the circuit breaker should let short-circuit current pass through it for a short period. Based on this discussion, it can be concluded that the rated short time current of a circuit breaker is equal to or greater than the rated short-circuit breaking current of the CB.

### 1.2.5 Rated voltage of circuit breaker

The insulation system is supposed to bear the electrostatic stresses due to operating voltage levels, and hence, it decides the rated voltage of circuit breaker. For any operating voltage below 400 kV, the over voltage limit is taken 10% above the rated voltage. Also, every CB is supposed to withstand slight over voltages. For voltage levels above 400 kV, it is 5%. These limits accommodate slight over voltages allowed during no load. A circuit breaker is also subjected to two other high voltage conditions:

(1) Due to Ferranti effect, if large loads are disconnected, it results in high induced voltages. This voltage which appears across the contacts of CB may be significantly higher than the highest system voltage. Frequency of this voltage is that of power and for a short period of time to be taken care by protection system. To address this, CB must be rated to withstand this voltage for a short time (generally 60 s). As any longer duration will make the cost of circuit breaker quite high and in normal case, the situation does not last more than 60 seconds.

(2) Another instance is due to nature, whenever, lightning strikes a line chances are that the high voltage surge may travel to breaker also. Thus, the insulation system and separation gap between contacts during open conditions should be capable of withstanding these two situations. This lightening surge is in the shape of peaky impulse and of microsecond duration. Therefore, the circuit breaker should be capable of withstanding it. For different system voltages, the related information is shown in Table 10.2.

**TABLE 10.2** Different values of system voltage along with power frequency and impulse voltage level [3].

| System voltage (kV) | Highest system voltage (kV) | Power frequency withstand voltage (kV) | Impulse voltage level (kV) |
|---|---|---|---|
| 11 | 12 | – | – |
| 33 | 36 | 70 | 170 |
| 132 | 145 | 275 | 650 |
| 220 | 245 | 460 | 1050 |
| 400 | 420 | – | – |

## 1.3 Condition monitoring of circuit breaker

High-voltage circuit breakers (HVCBs) are considered as important elements because of the duty assigned to them. Thus, reliability and resilience of power system is significantly affected. Many failures, including explosions, have been reported due to non-perfect operation of the CB. As has already been discussed, CBs consist of various mechanical and electrical components such as dampers, latches, springs, and coils, thus, resulting in failures emerging from different sources. Therefore, proper maintenance policies can prevent the failures and extend the operational life of CBs [4,5].

Majorly, CBs have three main sub-sections: the control section, operating mechanism and interruption chamber. Major classification of circuit breakers is based on the arc quenching medium. For arc quenching aspects, circuit breakers are air, minimum oil, SF6 and vacuum CBs. In general, SF6 CBs are used in high-voltage applications, whereas vacuum CBs are preferred for LV and MV power networks due to certain restrictions. Recent research has focused on finding alternatives to SF6 to mitigate environmental impact. Due to recent advancements in renewable power generation and HVDC system, attempts are being made to use power electronics in circuit breakers, which are known as hybrid circuit breakers. These hybrid CBs enable efficient current interruption within acceptable operation times.

Inspection and maintenance play a crucial role in extending the lifetime of CBs and preventing failures. Various maintenance schedules are utilized, including time-of service-based maintenance (TBM), another is based on situation warrants, that is, condition-based maintenance (CBM), and lastly reliability-centred maintenance (RCM). TBM, also known as proactive maintenance, involves performing predefined activities at regular intervals to prevent component failures. However, this approach may not consider the condition of the components, leading to potential issues if a serious problem arises before the next inspection. In some cases, TBM may result in increased costs without necessarily improving reliability [6].

CBM, on the other hand, focusses on determining the inspection intervals and activities based on the condition of the CBs. Some routine operating data is used for diagnostic purposes such as cycles of operations, operating coil current (CC), timing tests, etc. This information is used to assess the condition of circuit breaker elements. CBM is also referred to as predictive maintenance. RCM involves implementing CBM strategies based on the location and impact on system of CBs in power systems and allocating the maintenance budget accordingly. This approach ensures that maintenance efforts are directed towards critical parts of the network.

Among the various maintenance schedules, CBM is more applicable with the future smart environment due to its use of diagnostic parameters. Understanding failure modes, the origins of failures, and utilizing appropriate captured signals are crucial steps in implementing maintenance strategies for HVCBs.

## 1.4 Monitoring parameters

A wide range of captured signals can be utilized to evaluate the condition of circuit breakers (CBs) in both offline and online scenarios. In this context, it is essential to emphasize the significance of key diagnostic signals that provide valuable insights into health of circuit breaker, the challenges faced. Summary of measured parameters and monitored functions/components are provided in Table 10.3.

### *1.4.1 Coil current*

Fig. 10.1 shows major components of a typical 72.5 kV, $SF_6$-CB [1,2].

A typical measurement data for trip coil current and its diagnostic features are shown in Fig. 10.2. As shown in the figure current, the current profile is characterized by time- and current-based features [3–5]. When an abnormality in system occurs requiring operation of circuit breaker, protection system issues a trip signal to circuit breaker by energizing trip/close coil. When trip coil current reaches operating value (IP1 at TP1), the force is good enough to move armature in the coil to hit the latch (at TV). This movement of armature causes the release of charged operating mechanism (usually the charged spring(s)) to start the motion of contacts through the mechanical transmission linkage (TP2, IP2).

Sincere work is performed to establish a relation between captured signal and diagnosis therefrom. These signals are also used in fault detection algorithms. Efforts are also made to find correlations among various captured parameters. Further efforts have been made to establish relationships among

**TABLE 10.3** Summary of measured parameters and monitored functions/components [4].

| | Measured parameter | | | | | | | | | |
|---|---|---|---|---|---|---|---|---|---|---|
| Monitored-function/ component | Contact travel | Coil voltage | Coil current | Motor voltage | Motor current | Aux contacts | Main current | SF6 density | SF6 humidity | Cabinet temperature |
| **Operating mechanism** | | | | | | | | | | |
| Operating time | | x | x | | | x | | | | |
| Velocity | x | | x | | | x | | | | |
| Over travel | x | | | | | | | | | |
| Charging time | | | | x | x | | | | | |
| Charging current | | | | | x | | | | | |
| Number of charges | | | | x | x | | | | | |
| Energy stored | | | | x | x | | | | | |
| **Main circuit** | | | | | | | | | | |
| Operating time | x | x | x | | | x | | | | |
| Operating speed | x | | x | | | x | | | | |
| SF6 | | | | | | | | x | x | |
| Cumulated interruptions | x | | x | | | x | x | | | |

*Continued*

**TABLE 10.3** Summary of measured parameters and monitored functions/components [4].—cont'd

| Monitored-function/ component | Measured parameter | | | | | | | | | |
|---|---|---|---|---|---|---|---|---|---|---|
| | Contact travel | Coil voltage | Coil current | Motor voltage | Motor current | Aux contacts | Main current | SF6 density | SF6 humidity | Cabinet temperature |
| Insulation | x | | | | | x | | x | x | |
| **Auxiliary circuits** | | | | | | | | | | |
| Battery voltage | | x | | x | | | | | | |
| Coils integrity | | x | | | | | | | | |
| Temperature | | | | | | | | | | x |
| **Tightness** | | | | | | | | | | |
| Leakage | | | | | | | | x | x | |

❶ Travel Curve Transducer
❷ Main shaft

**FIGURE 10.1** Summary of measured parameters and monitored functions/components. It shows major components of a typical 72.5 kV, SF6-CB [1,2].

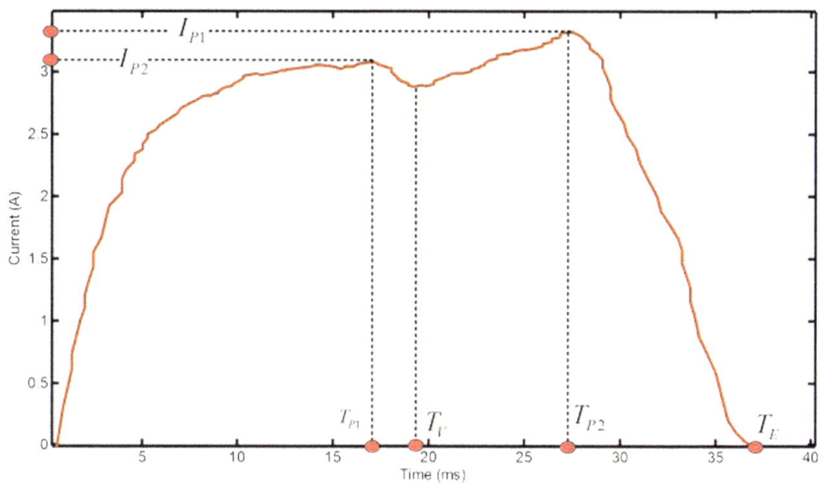


**FIGURE 10.2** Trip coil current w.r.t time.

multiple failures involving the latch, control voltage, and coil. Magnitude and nature of the coil current change based on health of the operating mechanism. Thus, it can be used to predict faults. This signal in conjunction with the contact travel curve (TC) has been utilized for fault detection in HVCBs. Conditions such as degraded battery, partially damaged contacts, and coil and friction have been mapped using CC. The CC signal is also effectively used for ageing of coil, core jamming, and lack of core travel space. Furthermore, timing features of CC signal can be utilized to improve fault detection based on the vibration signals. Data mining approach can be used to fix threshold values for various feature parameters to diagnose the fault. In this case, a probabilistic approach is used.

The coil current can be a good reference for failure detection in trip/close coil supply voltage, continuity of the coil, latch, and auxiliary switches. Moreover, the poor health of station battery results in voltage applied on the control unit of CB can be lower or higher than the set value. With age of coil, despite continuity being maintained, resistance varies and this can be effectively used for assessing age of coil and its corresponding probability of malfunction. This failure has an influence on the CC profile. Furthermore, if movable mechanical parts are not properly lubricated, the latch operation can be imperfect due to deterioration of lubricating property.

Signals obtained from different elements not only provide real-time information about the health of the CB as a whole and of the operating mechanism in specific. It is worth mentioning that arrangements required for these measurements are quite simple and easy to implement. For this purpose, a Hall effect current sensor or resistance in shunt is to be deployed which is also inexpensive. The CC provides only the information pertaining to failures of control and auxiliary circuit. Thus, merely this is not enough for complete assessment of condition of CBs as it can only be used to identify only 25% of faults.

### 1.4.2 Travel curve

Travel curve (TC) stands for timing characteristic, or in other words, it provides information about displacement of contacts as a function of time. It is measured using a transducer and is illustrated in Fig. 10.3 [3–5].

The TC represents the close–open stroke–time characteristic of CB. The operation speed of the CB can be determined by calculating the time between

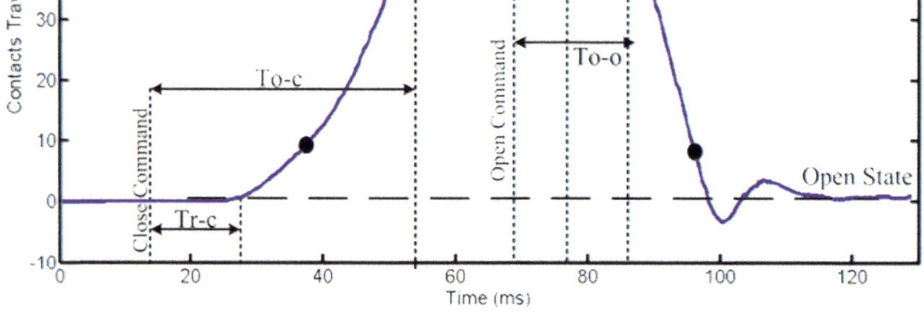


**FIGURE 10.3** Timing characteristic.

two prefixed points on this curve. Additionally, timing values such as the reaction time in closing/opening (Tr) and mechanism time (To) can be obtained from the TC plot. This information is of prime importance for evaluating the operation time of CBs through timing tests and identifying malfunctions in the operating mechanism.

Analysing the TC can help establish a correlation between failures and diagnostic features. This correlation is crucial in developing intelligent fault detection algorithms. This correlation of TC profile can be used to detect the problems in latch supply voltage, coil defects, damper damages, leakage/spring settings, contact issues as well as that of pull rod. Changes in supply voltage can affect the operation speed of the trip/close coil for this timing characteristics TC can be used. Different segments of TC can be used to identify different abnormalities. The tail of TC curve provides information regarding anomalies in dampers. Improper driving mechanism for spring, settings as well as leakage of oil in the hydraulic driving mechanism can cause variations in the speed and timing characteristics of TC. Therefore, any malfunction in driving mechanics is reflected in TC.

The main advantage of TC is its capability to provide crucial information related to failures originating from the operating mechanism of CBs. In addition, TC significantly affects the dynamic contact resistance. Hence these two combined together can be used for assessing health of both sets of contacts. This information is typically available in the data sheet provided by manufacturers of CBs as part of routine tests, making it a useful tool for condition assessment. One disadvantage of using TC is that it requires the installation of a transducer sensor, which limits its applicability for online assessments in certain CB designs.

### *1.4.3 Dynamic resistance measurement*

During the operation of CBs under normal and short-circuit currents, both the main and arc contacts can undergo deterioration. To assess the condition of these contacts without the need of dismantling the CBs, dynamic contact resistance measurement (DRM) is a standard non-invasive method to evaluate the health of contacts along with the operating mechanism. DRM is conducted by measuring voltage across contacts under opening operation, when a known value of DC current is injected. This allows for the determination of the variation of resistance as a function of the contact displacement as shown in Fig. 10.4. For this purpose, moderately high value of DC current in the order of 200-A is injected into the CB. As the movable contacts of CB starts moving, the lengthening of the current path leads to an increase in dynamic resistance. During the breaking operation initially, main contacts are in picture and later auxiliary contacts come into picture. Hence, initial part of the resistance profile represents resistance of main contact. Due to higher resistance of material used for auxiliary contacts, a significant change in DRM is obtained when the

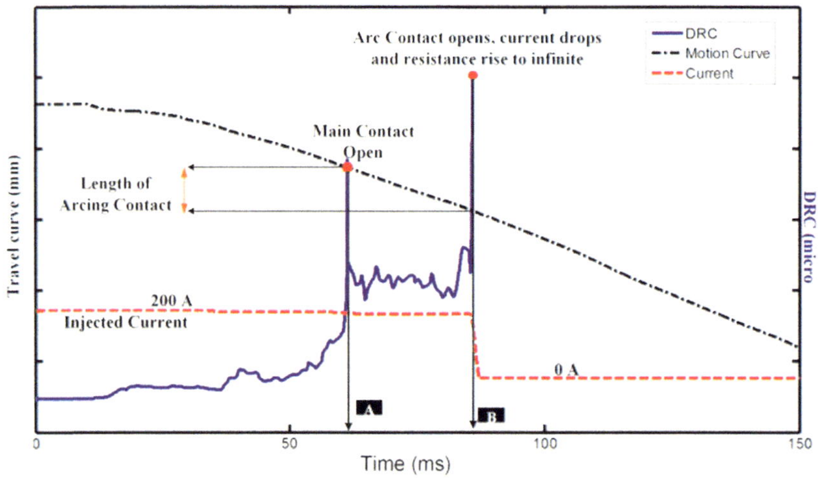


**FIGURE 10.4** Travel curve.

current is transferred from main contacts to auxiliary/arcing contacts. It may show fast variation (spikes) due to cyclic operations, which results in heating the contact material, thereby deteriorating its mechanical property. After a certain duration, the current is interrupted. By analysing the combination of the motion curve and DRM, it is possible to obtain information about the main, arc contact resistances and the length of the arc contact.

If either of the main or arcing contacts is eroded, the DRM profile will exhibit changes compared with that in healthy condition. Erosion increases the resistance, while shortening affects the separation time. Therefore, malfunctions in the contacts can be identified through DRM analysis without the need for dismantling the CBs. It is important to note that establishing a proper correlation between failures and the diagnostic features provided by DRM is very crucial in the diagnosis process. Studies have been conducted to simulate the quantum and its impacts of erosion for fixed and moving both arcing and main contacts, as well as under broken fingers on DRM profiles. However, in the initial stages of erosions, it will be difficult to assess exact quantum of erosion. It primarily depends upon the experience of the engineer. Yet another issue is that studies have proven that dynamic resistance curve fails to reflect beyond doubts the early stage of contact erosion. Nevertheless, contact erosion can still be calculated using diagnostic features obtained from the DRM profile.

Real advantage of DRM is that the health of arcing and main contacts can be obtained without the need of dismantling the CB. The serious limitation being it can be used only in offline mode, i.e. the breaker has to be withdrawn from service. The arc contact length assessment requires TC; thus, results are highly dependent on magnitude of injected current as well as clarity of profile to operating speed of the CBs. The outcome of this technique and its repeatability depends on many factors. For example, the measured resistance through this test depends on opening velocity and the magnitude of injected current. This resistance has an inverse relation with the injected current and direct relation with velocity of contact opening.

### 1.4.4 Energy

As each operation of CB results in oxidation of some contact material, it can be used as a parameter to find residual life of CB. Recently, loss of contact mass through the energy dissipated during making and breaking operation is used for residual life estimation of CB. It is also used for assessing the health of contacts. Real-time applicability of this approach gives it an edge and can be considered as an online version of dynamic resistance measurements (DRMs).

Another novel approach is to design a circuit that can measure arc voltage in real time. This voltage can be used to predict residual life of CB based on energy. The energy of arc is calculated using the arc current recorded through a current transformer/current probe, and the arc voltage measured by deploying the suggested online measurement system. However, using energy as an assessment method for estimating interrupter wear has its challenges and disadvantages. The accuracy is limited due to the omission of voltage information. Due to the presence of high-voltage transients, it poses a challenge to measure the arc voltage with precision. For this purpose, special tools are required.

### 1.4.5 Vibration

The analysis of vibration signals produced while opening and closing of CB has gained popularity in recent years due to convenience in data acquisition and suitability for non-invasive and real-time evaluation. This method allows for the detection of operating mechanism of CBs failures, covering various potential sources of issues.

The vibration signal, as shown in Fig. 10.5, can provide insights into anomalies in the mechanical section of the CBs [7,8]. When failures occur, in the damper, rod, or shaft, the vibration signal gets changed from those that of when healthy. Health conditions involving deformation of the shaft, close coil, insulation pull rod, and damper can be investigated using vibration signal analysis. To minimize the arcing duration, the contacts are very rapidly moved, and to absorb the residual energy, post-operation dampers are used. Thus, damper is responsible for absorbing residual energy of spring and reducing the impact of mechanical collisions during the operation of HVCBs. The insulation push/pull rod passes the energy from the operating mechanism to the moving contacts for opening and closing operations. Fig. 10.5 gives vibration burst resulting from a failure in the oil damper, as well as the vibration signal

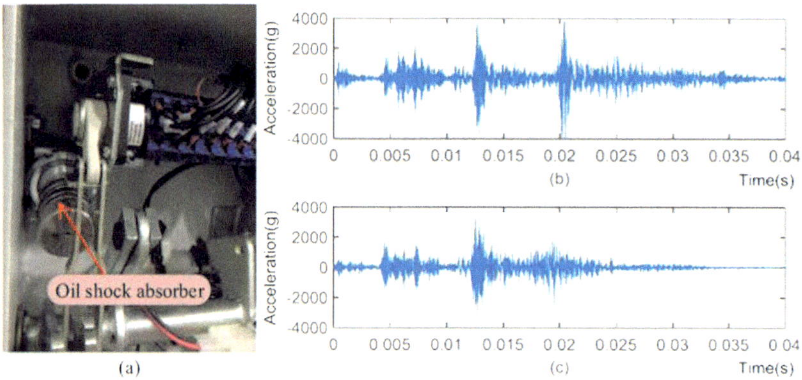


**FIGURE 10.5** Vibration signal.

when healthy. It is to be noted that the vibration signal spectrum changes when a fault occurs. It is worth mentioning that post-processing techniques are required for discriminating between healthy and faulty vibration signals.

A post-analysis of the vibration signal provides real-time assessment of potential failures originating from various parts of the operating mechanism. Thus, maintenance can be done proactively. Moreover, it includes all the vibrations caused by all mechanical parts as well as electromechanical vibrations. As the vibration signal is for a short duration, its non-linearity makes the sensor selection and its placement very crucial to make this method acceptable. Further, the algorithm is used to establish correlation of this signal with probable cause of failure. The signal is non-stationary and contains wide-frequency spectrum.

### *1.4.6 Environmental requirements*

It is crucial for a monitoring system in a high-voltage substation to operate reliably, even in challenging environments with high electromagnetic disturbances. A primary requirement for such a system is high electromagnetic immunity, ensuring it can withstand both radiated and conducted energy. Additionally, the working temperature range of the monitoring system should be taken into consideration. In many cases, the instrumentation system required for monitoring environmental parameters is usually installed inside the control cabinet of the CB [5]. This system is required to be protected from excessively as well as ingress of sand if located in deserts; where temperatures may be of the order of 60°C, ventilation is not recommended. Due to lack of ventilation, temperature may rise well above 80°C. This high temperature is a matter of concern and is addressed by choosing devices which can operate up to 85°C without any deterioration in performance. Generally, the cubicle temperature and atmospheric temperatures monitored and logged at regular interval are used to correlate alarms of SF6 and for working time heaters [6].

### 1.4.7 Insulating/quenching medium

For HVCBs, one important parameter required to be monitored is the density or pressure of $SF_6$ gas used as an insulating medium [7–9]. Measuring the $SF_6$ density or pressure, rescaled to 20°C, serves two main purposes:

1. Prevent or limit leakage: Monitoring $SF_6$ density helps to detect any leakage, which is crucial for environmental reasons. Maintaining the integrity of the $SF_6$ gas system ensures that leakage is minimized, reducing the impact on the environment [8,9].
2. Ensure breaker functionality: $SF_6$ gas is essential for insulation and effective switching to clear electrical faults. Monitoring $SF_6$ density is crucial to ensure that the breaker operates within the desired range. $SF_6$ density relays are commonly installed in circuit breakers, with pre-set alarm and lockout levels. If the density falls below the lockout level, the breaker cannot be operated [9,10].

Using an online gas density metres along with the density relays can be used for trend analysis which leads to early detection of any $SF_6$ leakage (as depicted in Fig. 10.6). This monitoring system is used to send alarm/warning for $SF_6$ leakage whenever pressure falls below threshold. Moreover, the logged density data can be used to forecast critical level will be reached. Hence, it is required to provide valuable information to asset managers in planning for maintenance or inspection. This entire process is fully automated [9,10].

Fig. 10.6 illustrates the comparison between $SF_6$ monitoring and relay gas pressure detection. Trend of these parameters can be used to early detection of leaks as well as when the alarm and lockout pressure will be reached. The density log can be accessed remotely by the asset manager, allowing them to make informed decisions regarding maintenance planning.

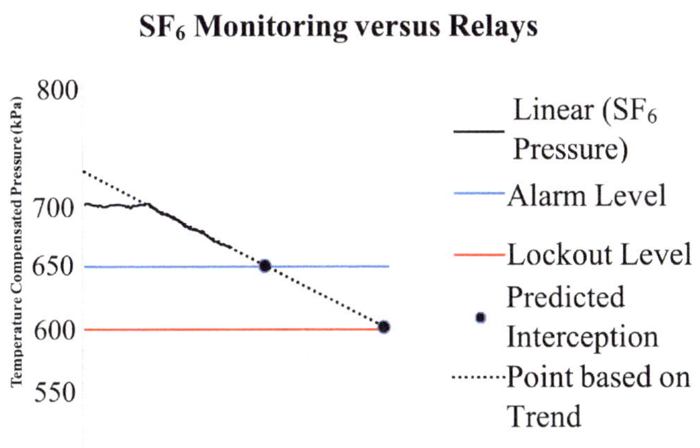


FIGURE 10.6 Comparison between SF6 monitoring and relay pressure detection.

In addition to monitoring the quantity of $SF_6$, 'smart' sensors also enable the assessment of its quality by measuring the moisture content [8,10]. These sensors can be connected through a Modbus RS-485 communication port. Early detection of critical trends through continuous monitoring allows for proactive maintenance planning instead of reacting to urgent situations, as in the case of periodic $SF_6$ analysis.

### 1.4.8 Operation and timing

For every cycle of operation, an event is recorded, and the operating time for each closing and opening operation is measured. These measured values are then compared with the reference values that were recorded as a 'fingerprint' during the breaker's installation or after maintenance. Several parameters are measured during the operation of the circuit breaker [9]:

1. Voltage command: The magnitude of voltage applied to the control circuit can be used to discriminate among different operating commands.
2. Current in the opening and closing coil: Measuring the current flow in the coil ensures that the operating command has been received. This is important as the current flows through all the interlocking relays of the control circuit. Information about the health of operating coil can be obtained from the shape of the coil current. Simultaneous measurement of voltage and coil current provide critical inputs for the diagnostic capability, especially in erroneous wiring or coil.
3. Commutation of auxiliary contacts (52 a/b): During the operation of CB, the time taken for shifting from main contact to auxiliary contact and vice versa is measured. The difference in time between the auxiliary and main contact separation is recorded in the test report and is used to evaluate the correct operating time.

To ensure that the measurement does not alter the closing and opening operating circuits, common practice is to measure the coil current using a non-invasive current sensor. For AC coils, a split-core current transformer is typically used but modern CBs use current probe, while for DC coils, a Hall effect sensor is preferred over a shunt. Using a Hall effect sensor eliminates the need to insert a shunt in series within the DC control circuit, reducing the potential for failure. It is important to establish reference 'fingerprint' values for both the current impulse and the commutation of auxiliary contacts, as they serve as a baseline for comparison during subsequent operations and diagnostics. By analysing the measured values, potential issues or deviations from the reference values can be identified, enabling early detection of faults, and facilitating maintenance planning.

### 1.4.9 Auxiliary voltage

Monitoring the battery voltage can serve as a diagnostic help, especially in cases of failure to operate situation. Certain threshold voltage of battery is required for proper operation. Generally, the trip circuit requires the battery voltage to be above 70% of the nominal value, while the closing circuit should be above 85% of the nominal value. These thresholds ensure that the operating coils receive sufficient voltage to perform their intended function reliably. Regularly monitoring and maintaining the battery voltage at appropriate levels is essential for ensuring the proper operation and functionality of the circuit breaker. Any deviations in the battery voltage below the specified thresholds can be indicative of potential issues or the need for battery replacement or maintenance.

### 1.4.10 Charging motor of the operating mechanism

For the operation of CB, an operating mechanism is provided in every circuit breaker. The purpose is to provide necessary energy for its operation. In the case of spring, energy is stored in the form of elastic energy, while for hydraulic operating mechanism, it is stored as fluid pressure. Thus, after each operation, it is crucial to restore this stored energy. To this end, an electric motor is used to charge the spring or pump fluid in case of hydraulic operating mechanisms.

The duration of charging and the value of charging current of the operating mechanism can be used to gather valuable information about its health condition. For the measurement of charging duration, the presence or absence of voltage applied to the motor. Additionally, measuring the absorbed current confirms that the motor is running and functioning as intended.

Changes in the absorbed current of charging motor can serve as a warning for possible problems in the operating mechanism [9]:

1. Higher current values: Higher current can be attributed to either increased friction in charging mechanism or an over-range of the closing spring. This could be due to a mis-adjustment of the end switch, which fails to stop the motor at the charged spring position.
2. Lower current values: A decrease in absorbed current could also be an indication of a mis-adjustment of the end switch or reduced tension of the closing spring. Confirmation of this can be made by recording a small over travel during closing.

Frequent recharge and longer recharge duration without operation and longer operation of recharge motor after operation in hydraulic drives can be attributed to a possible oil bleeding from the high-pressure compartment to the low-pressure chamber. Monitoring these parameters can help identify potential issues with oil leakage. Monitoring the charging time, charging current and changes in absorbed current provides important insights into the health and functionality of the operating mechanism. Any deviations from expected values can indicate potential problems that may require adjustment or maintenance to ensure the proper operation of the circuit breaker.

### 1.4.11 Main current and electrical wear

The amount of wear will depend on the amount of heat produced which depends on the amount of current interrupted. Thus, measuring the interrupted current for each operation can be used to estimate the electrical wear of the CB contacts. The arcing contact erosion and nozzle ablation occur during the interruption process when the arc is burning between the contacts of the CB. The commonly used 'integral square t' formula is a more general formula, as given by Eq. (10.2).

$$wear \approx K(i) \int_{i=0}^{CS} |i|^n dt \tag{10.2}$$

where the exponent 'n' is a design-dependent parameter that should be provided by the CB manufacturer, typically ranging between 1 and 2. $K(i)$ represent the wear coefficient function of current level, which takes into account the different effectiveness of wearing processes at different current levels. It is also specific to the CB design. As the arc starts at the point of separation of contacts and continues till current reaches natural zero, the integration in the formula should be carried out for this duration only.

Commutation of auxiliary contacts (52 a, b) along with an offset trip factor in milliseconds is used to find contact separation time. This information is made available in the routine test report of CB. Similarly, the integral of opening time from every open operation determines the accumulated wear effect. These values can be compared with the benchmark values provided by the manufacturer to assess wear effect.

For current measurement, an auxiliary current transformer is connected to the secondary winding of the main current transformer installed in the substation. The metering core is required to measure current during normal operation which may be up to 120% of the rated value (to avoid over loading of metering instruments) but need to be quite accurate, whereas protection core can be subjected to as high as 10 times the rated current so it should have very high saturation level although high accuracy is not expected.

It is not recommended to use invasive methods for its measurement due to secondary winding of the core of a current transformer being part of the protection circuit. To this end, a common practice is to use non-invasive measurement transducer, such as a coaxial split-core auxiliary current transformer that can be clamped around the wire. This results in convenient installation and non-interference with protection circuit [6,9].

## 1.5 CB failure rate

The proper functioning and reliability of power CBs rely heavily on the capabilities of their components, control systems, and operating mechanisms. The components of a CB can be categorized into three main parts: live parts and insulation, control parts and operating mechanism parts [6,9].

Firstly, the live parts and insulation are further divided into two sub-components. The first sub-component is the main insulation, which provides insulation to the earth. The second sub-component is the making and breaking unit, which is responsible for the opening and closing of the CB.

Secondly, the control parts of a circuit breaker encompass various elements including auxiliary switches, associated drives, contractors, relays, heaters, thermostats, fuses, gas density supervision, and tripping and closing units.

Lastly, the operating mechanisms of the CB consist of several components. These include the actuator, compressors, motors, pumps, pipe unions, control elements, counters, damping devices, energy storage systems, mechanical transmissions and non-return devices. It is crucial to ensure the proper functioning and coordination of all these components, control systems and operating mechanisms to maintain the performance and reliability of power circuit breakers. Historical data is illustrated in Fig. 10.7 and Table 10.4 as per investigation done by CIGRE [1,2].

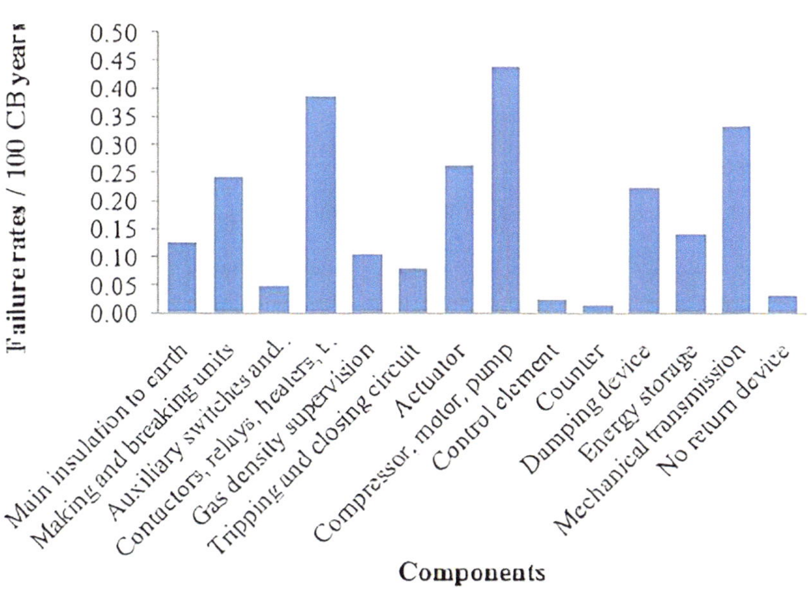


FIGURE 10.7 Failure rate of different components of CB.

**TABLE 10.4 Historical data for failure rate of different components of CB.**

| Type of components | No. of failures | CB years | No. of failures/ 100CB years |
|---|---|---|---|
| Main insulation to the earth | 31 | 24472 | 0.127 |
| Making and breaking units | 60 | 24472 | 0.245 |
| Auxiliary switches and associate drives | 12 | 24472 | 0.049 |
| Contractors, relays, heaters, etc. | 95 | 24472 | 0.388 |
| Gas density supervision | 26 | 24472 | 0.106 |
| Tripping and closing circuits | 20 | 24472 | 0.082 |
| Actuator | 65 | 24472 | 0.266 |
| Compressors, motors, pumps, pipe union | 108 | 24472 | 0.441 |
| Control elements | 6 | 24472 | 0.025 |
| Counter | 4 | 24472 | 0.016 |
| Damping device | 55 | 24472 | 0.225 |
| Energy storage | 35 | 24472 | 0.143 |
| Mechanical transmissions | 82 | 24472 | 0.335 |
| Non-return device | 8 | 24472 | 0.033 |
| Total | 607 | | |

## References

[1] CIGRE A3-204, Practical Circuit Breaker Condition Monitoring with Protective Relays, 2012.

[2] C. Schneider, M. Skidmore, Z. Campbell, J. Byerly, K. Phillips, Circuit breaker asset management using intelligent electronic device (IED)-Based health monitoring, in: CIGRE US National Committee 2014 Grid of the Future Symposium, American Electric Power (AEP), United States, 2014.

[3] IEEE Std C37.10-1995 (R2002) IEEE Guide for Diagnostics and Failure Investigation of Power Circuit Breakers, 1995.

[4] P. Dehghanian, T. Popovic, M. Kezunovic, Circuit Breaker Operational Health Assessment via Condition Monitoring Data, Texas A&M University, Texas, 77843, USA, 2014.

[5] R. Arias, M. Runde, K. Takahashi, J.A. Wiersma, User Guide for the Application of Monitoring and Diagnostic Techniques for Switching Equipment for Rated Voltages of 72.5kV and above, TB 167 CIGRE WG 13-09 (Monitoring and Diagnostic Techniques for Switching Equipment), August, 2000.

[6] Final Report of the 2004–2007 International Inquirí on Reliability of High Voltage Equipment—Part 2-Reliability of High Voltage SF6 Circuit Breakers.

[7] The Indian Electricity Grid Code with Vide. No. L-1/265/2022/CERC, June 07, 2022. https://www.cercind.gov.in. https://cercind.gov.in/2022/draft_reg/Draft-IEGC-07062022.pdf.

[8] Z.H. Zargar, K.J. Akram, G.R. Biswal, T. Islam, A linear capacitive sensor for PPM moisture measurement in $SF_6$ gas insulated switchgear, IEEE Trans. Instrum. Meas. 70 (Jan. 2021) 1–8. Art. No. 9001608.

[9] B. Bhalja, R.P. Maheshwari, N.G. Chothani, Types of circuit breaker and their testing, in: Protection and Switchgear, first ed., Oxford Univ. Press, New Delhi, India, 2013, pp. 344–387 (Chapter 14).

[10] G.R. Biswal, T. Islam, S.D. Das, K.J. Akram, Ultrasonic Sensor Based Gas Density Monitoring of SF6 Gas Insulated Switchgear, Jan. 29, 2024. Indian Patent Application No.: 202111011524 A, Filing Date: Mar. 18, 2021; Publication Date: Apr. 16, 2021.

Chapter 11

# High-Voltage DC applications for switchgear

**Sweta Panda[1], Kalyan Dash[2] and Biswajit Tewari[3]**
[1]*Veer Surendra Sai University of Technology, Department of Electrical & Electronics Engineering, Burla, Odisha, India;* [2]*Continuum Associates LLC, Millburn, NJ, United States;* [3]*Electrical & Instrumentation, Hindalco Industries Ltd., Hirakud, Odisha, India*

## 1. Introduction

The evolution of electrical power transmission has witnessed a transformative shift with the advent of high-voltage direct current (HVDC) technology. As demands for efficient, long-distance power transmission and the integration of renewable energy sources (RESs) continue to rise, HVDC has emerged as a crucial component in modern power systems. A key advantage is that transmission loss is less with HVDC compared with alternating current (AC) systems, making it ideal for long-distance transmission. HVDC's reduced energy losses benefit the integration of remote renewable sources, ensuring more power reaches end users efficiently. This chapter delves into the intricate world of HVDC applications for switchgear, exploring the fundamental principles, challenges, and specialized components that define the landscape of HVDC transmission.

### 1.1 Overview of high-voltage DC technology

HVDC technology has emerged as a revolutionary force in the realm of electrical power transmission. Rooted in decades of technological evolution, HVDC systems operate on the concept of converting AC to direct current (DC) and then exchanging power between sending and receiving end, a paradigm shift from traditional HVAC systems. This technology has gained prominence for its efficiency in transmitting power over extended distances, overcoming the limitations associated with HVAC systems and providing a versatile solution for the modern energy landscape.

At the core of HVDC technology lies the fundamental principle of regulated conversion between AC and DC, enabling controlled power transmission

**Switchgear Design, Operation, and Maintenance Using Industry Standards**
**https://doi.org/10.1016/B978-0-443-24718-7.00011-1**

over vast distances. The advantages of HVDC are multifaceted, enhancing grid stability, facilitating long-distance power exchange and enabling seamless integration of RESs. HVDC's proficiency in long-distance power transfer, coupled with superior stability and control, makes it an ideal solution for cross-border power transmission and offshore wind farm connections. Additionally, HVDC's capability to interconnect asynchronous sources and integrate renewable energy positions it as a critical asset in the quest for a more reliable, flexible and sustainable power infrastructure. This technology is a cornerstone in global efforts to transition towards a more resilient and environmentally friendly energy system.

As technology advances, ongoing research focusses on improving converter efficiency, refining control systems and exploring innovative materials. The integration of digital technologies, such as advanced monitoring and control systems, is further propelling HVDC towards a future characterized by enhanced efficiency, reliability and adaptability to the evolving energy landscape. The ongoing developments and future trends underscore the dynamic nature of HVDC technology, positioning it as a crucial element in the pursuit of a sustainable and technologically advanced energy sector.

## 1.2 Growing importance of HVDC in modern power systems

The increasing importance of HVDC technology in modern power systems is a testament to its transformative impact on the energy landscape. Unlike traditional HVAC systems, HVDC has become a linchpin in addressing contemporary challenges and meeting the evolving needs of power transmission.

One of the primary factors contributing to the growing significance of HVDC is its unparalleled efficiency in long-distance power transmission. As demand for energy rises and RESs become distributed, HVDC's ability to transmit power over extended distances with reduced losses makes it a reliable and effective alternative for cross-border power transmission.

The higher controllability of HVDC system enables stable operation even with the intermittent renewable sources such as PV and wind. Seamless integration of these asynchronous sources contributes towards the goal of sustainable energy systems. Furthermore, as power grids continue to expand and evolve, HVDC technology offers a solution for reinforcing existing AC grids. By alleviating congestion and enhancing overall grid reliability, HVDC contributes to a more resilient power infrastructure.

Thus, the growing importance of HVDC in modern power systems is a result of its unmatched efficiency, adaptability to renewable energy integration and capacity to reinforce existing power grid. As the energy landscape continues to evolve, HVDC technology stands as a cornerstone alternative, facilitating the transition towards a sustainable, reliable and interconnected global power infrastructure.

## 1.3 Need for specialized switchgear in HVDC application

The integration of HVDC technology into modern power systems brings forth a distinctive set of challenges and operational requirements, necessitating the development and deployment of specialized switchgears. The transition from traditional HVAC to HVDC transmission underscores the critical role that purpose-built switchgears play in ensuring the reliability, safety and efficiency of the system.

One of the primary drivers behind the need for specialized HVDC switchgear is the fundamental difference in the characteristics of DC and AC systems. Unlike AC system, which contains a sinusoidal current with zero crossing, DC system has a continuous current waveform. This poses unique challenges for interrupting and isolating transmission lines during fault condition, demanding innovative solutions that go beyond the capabilities of conventional AC switchgear.

HVDC switchgear must address the specific requirement of converter stations, the prime component in HVDC systems responsible for AC to DC conversion and vice versa. These converter stations involve complex processes such as voltage conversion, filtering and synchronization that require specialized switchgears to handle the unique demands associated with these functions. The switchgears must exhibit high reliability and performance to ensure the seamless operation of converter stations under varying conditions.

Additionally, insulation system in HVDC application needs to contend with challenges which are not present in case of AC system. The absence of natural current zero crossing in DC system may result in a sustained arc necessitating advanced insulation material and design capable of withstanding continuous arcing conditions. Specialized insulation solutions are crucial for preventing breakdowns, ensuring safety and integrity of personnel and equipment.

Control and protection systems represent another critical aspect of HVDC switchgear. The precise control of power flow and the implementation of rapid and accurate protection mechanisms are essential for maintaining grid stability and preventing catastrophic failures. Specialized control and protection devices are designed to handle the intricacies of HVDC systems, responding promptly to faults and ensuring the secure operation of the overall network.

Hence, the need for specialized switchgear in HVDC applications is driven by the unique characteristics and operational challenges posed by DC transmission. Purpose-built switchgear is essential for addressing the intricacies of converter stations, adapting to the continuous nature of DC systems, providing advanced insulation solutions and ensuring precise control and protection. The development and deployment of specialized HVDC switchgear are pivotal steps in unlocking the full potential of HVDC technology and fostering the evolution of resilient, efficient and secure power systems.

## 2. Fundamentals of high-voltage DC

The fundamentals of HVDC lay the groundwork for understanding the intricacies of this transformative technology, essential for efficient and reliable power transmission. This section explores the basic principles, crucial components and distinctive features that distinguish HVDC from traditional HVAC systems.

### 2.1 Basic principles of HVDC transmission

At the heart of HVDC transmission is the controlled conversion between AC and DC. This process is fundamental to efficient power transmission over extended distances. By converting AC to DC for transmission and vice versa at the receiving end, HVDC mitigates transmission losses, making it a preferred choice for long-distance power transfer. This subsection explores the principles governing HVDC transmission, providing insight into the mechanics of this technology.

#### *2.1.1 Comparison with high-voltage AC systems*

A critical facet of understanding HVDC technology is examining its distinctions from traditional HVAC systems. This comparative analysis provides insights into the unique characteristics and advantages that make HVDC an increasingly preferred choice in modern power transmission.

- **Efficiency in long-distance transmission:** HVDC exhibits superior efficiency over extended distances compared to HVAC systems. While HVAC systems experience significant transmission losses due to the inherent impedance and capacitance, HVDC minimizes these losses by maintaining a continuous and controlled flow of direct current. This characteristic renders HVDC especially advantageous for transmitting power over intercontinental distances.
- **Reduced reactive power components:** Unlike HVAC system that requires a large amount of reactive power, HVDC systems operate with minimal reactive power requirements. This reduction in reactive power leads to higher utilization of the transmission line, thereby increasing the power transfer capacity, beneficial for long-distance transmission.
- **Interconnection of asynchronous grids:** One of the distinctive features of HVDC is its ability to interconnect asynchronous grids operating at different frequencies. Traditional HVAC systems face challenges in synchronizing grids with varying frequencies, limiting their capability to interconnect diverse regions. HVDC technology transcends this limitation, providing a seamless connectivity for interconnecting grids with disparate characteristics.

- **Reduced electromagnetic interference:** HVDC systems produce lower electromagnetic interference compared to HVAC systems. The absence of frequent zero crossings in DC systems results in reduced electromagnetic emissions, contributing to better quality power.
- **Enhanced control of power flow:** HVDC systems offer enhanced control over power flow, allowing dynamic adjustments for optimized energy distribution. This flexibility is instrumental in adapting to fluctuating demand, intermittency in RESs, and, HVAC systems, in contrast, exhibit limitations in precise control and adaptability.
- **Economic considerations:** While HVDC infrastructure may entail higher initial capital cost, the overall lifecycle cost analysis often favours HVDC for certain applications. The efficiency gains, reduced transmission losses and the ability to transmit power over longer distances contribute to the economic viability of HVDC systems in specific scenarios.

### *2.1.2 Unique challenges and considerations in HVDC applications*

HVDC technology, while offering numerous advantages, introduces unique challenges and considerations that require careful attention for the successful implementation and operation of HVDC systems. Understanding and addressing these challenges are essential to ensuring the reliability, safety and optimal performance of HVDC applications.

- **Continuous current flow:** Unlike HVAC systems, HVDC system has a continuous flow of DC. This continuous current flow presents challenges during fault conditions, as interrupting DC circuits can be more complex than interrupting AC circuits because of nonexistence of natural zero crossings. Specialized circuit-breaking solutions are necessary to address this effectively.
- **Absence of natural zero crossings:** The absence of natural zero crossings in DC systems poses challenges for interrupting current during switching events. In AC systems, current naturally crosses zero during each half cycle, facilitating easier interruption. HVDC system lacks this inherent feature, requiring innovative solutions for interrupting and isolating circuits during fault.
- **Complexity in fault handling:** Faults in HVDC systems demand intricate fault-handling mechanisms. Rapid detection, localization and isolation of faults are crucial to prevent cascading failures and maintain the integrity of the power transmission network. Developing sophisticated protection schemes and reliable fault detection algorithms is paramount in HVDC applications.
- **Harmonics and filters:** HVDC system introduces harmonics into the grid due to use of power electronic converters, necessitating the use of filters to

mitigate their impact. Harmonics can lead to undesirable effects such as increased losses, interference with communication systems and potential damage to connected equipment. Proper filter design and implementation are essential to maintaining power quality and system stability.

- **Converter station complexity:** HVDC system involves complex converter stations that perform the crucial function of converting AC to DC at sending end and vice versa at the receiving end. Designing and operating these converter stations demand specialized knowledge and expertise. Factors such as voltage conversion, filtering and synchronization require careful consideration to ensure the seamless operation of HVDC systems.
- **Economic considerations and high initial costs:** While HVDC systems offer long-term economic benefits, their implementation involves high capital cost. The construction of converter stations, specialized switchgear and other components contributes to the upfront expenses. However, the overall economic viability, considering reduced transmission losses and increased efficiency, often justifies the initial investment in specific applications.

## 2.2 Components of HVDC switchgear

This section provides an in-depth exploration of the key components that constitute HVDC switchgear.

### *2.2.1 Converter station*

Converter stations serve as the central hub for HVDC transmission. It plays a pivotal role in the conversion of electrical power between AC and DC during transmission and reception. The schematic diagram of an HVDC converter station is shown in Fig. 11.1.

These stations are complex systems that include several essential components, each contributing to the seamless operation of HVDC technology. The components are given in brief in the following.

- **Converter transformers:** At the heart of converter stations are converter transformers, designed to facilitate the conversion of AC to DC and vice versa. These transformers convert voltage levels as required for transmission or reception. Converter transformers are specially designed to handle the unique characteristics of HVDC systems, such as continuous current flow and the absence of natural zero crossings.
- **Converters:** Converters are key components responsible for rectifying AC to DC at the sending end and inverting it back into AC at the receiving end. Modern HVDC systems often employ high-power semiconductor devices, such as thyristors or new gate injection transistors (GITs), in the

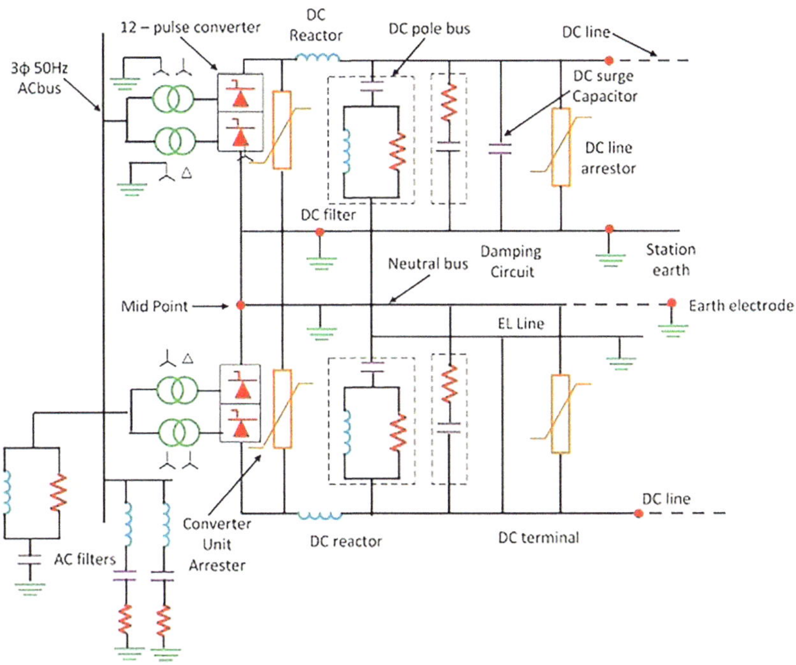


**FIGURE 11.1** Schematic layout of an HVDC converter station [1].

converters. The choice of converter technology depends on factors such as voltage levels, power ratings and system requirements.

- **Filters:** Filters are essential for maintaining the quality of power transmitted through HVDC systems. They are employed to mitigate harmonic distortions generated during the conversion process. By suppressing unwanted frequencies, filters contribute to the reduction of electromagnetic interference and ensure compliance with power quality standards. Harmonic filters, smoothing reactors and capacitor banks are the common types used in HVDC converter stations.
- **Valve halls:** In the case of line-commutated converters (LCC), converter stations include valve halls where the thyristor valves are housed. The thyristor valves control the flow of DC current and are crucial for the rectification and inversion processes. Proper cooling and insulation systems are implemented in valve halls to manage the heat generated by the thyristors during operation.
- **Control and monitoring systems:** Converter stations incorporate sophisticated control and monitoring systems to ensure the precise and stable operation of the HVDC system. These systems manage power flow, voltage regulation and converter operation. Advanced digital control technologies,

including microprocessor-based controllers, provide real-time monitoring, fault detection and adaptive control strategies.

Converter stations serve as the backbone of HVDC systems, facilitating the bidirectional flow of power between asynchronous grids or connecting RESs to the main power grid. The integration of advanced technologies within converter stations is essential for achieving high efficiency, reliability and control in HVDC transmission. Careful design, engineering and maintenance of converter stations are imperative for harnessing the full potential of HVDC technology in modern power systems.

### 2.2.2 Insulation systems

Insulation systems within HVDC switchgear are critical components designed to withstand the unique challenges posed by continuous DC transmission. They are essential for safeguarding the safety, reliability and efficiency of the entire HVDC system. This section delves into the key aspects of insulation systems in HVDC applications.

- **Continuous DC stress:** Insulation systems must be engineered to withstand the continuous DC stress, preventing breakdowns and ensuring the reliability of the HVDC network.
- **Switching events and arcing conditions:** During switching events or fault conditions, HVDC systems may experience sustained arcing. Insulation systems must be designed to endure these arcing conditions and maintain their effectiveness. Specialized materials, such as composite insulation with high resistivity and thermal stability, are employed to prevent insulation breakdowns during arcing.
- **Innovative insulation materials:** Insulation systems in HVDC switchgear must take into account innovative materials to meet the unique requirements of continuous DC operation. Advanced polymers, composite materials and insulating oils with enhanced dielectric properties are utilized to withstand the stress and environmental conditions experienced in HVDC applications.
- **Specialized design considerations:** The design of insulation systems considers factors such as the electric field distribution, voltage levels and environmental conditions. Proper dimensioning of insulation structures and the arrangement of insulating materials are crucial for maintaining the integrity of the system. Additionally, the insulation design must account for factors such as pollution, humidity and temperature variations, ensuring consistent performance over a range of operating conditions.
- **Surge arresters and overvoltage protection:** Surge arresters are integrated into HVDC insulation systems to protect against overvoltage that may arise due to lightning strikes, switching operations or other transient events. These devices divert excess voltage to the ground, preventing

insulation breakdown and ensuring the insulation's effectiveness in isolating circuits during normal and abnormal conditions.

- **Diagnostic and monitoring techniques:** Monitoring the condition of insulation is essential for proactive maintenance and early detection of potential issues. Diagnostic techniques, including partial discharge monitoring, insulation resistance measurements and thermal imaging, are employed to assess the health of insulation system.

### 2.2.3 Control and protection system

Control and protection are integral aspects within HVDC switchgear, ensuring the precise operation, stability and safety. This section explores the key functions and components that constitute control and protection systems in HVDC applications.

- **Power flow control:** Control systems within HVDC switchgear are responsible for managing the power flow between converter stations, ensuring optimal utilization of the transmission capacity. These systems employ sophisticated algorithms for control of converter to achieve the desire voltage and active-reactive power levels.
- **Voltage and reactive power regulation:** Voltage is regulated at both ends of the transmission line to maintain synchronism with the grid and counteract fluctuations and variations in the power system. Reactive power regulation is also crucial to optimize power factor and maintain voltage level of the HVDC network.
- **Dynamic response to grid conditions:** Control system exhibit dynamic response capabilities, adapting to changes in grid conditions and load demand. Advanced control algorithms enable swift adjustment to varying operating conditions, allowing HVDC systems to contribute to grid stability and respond efficiently to disturbances or changes in power demand.
- **Fault detection and handling:** Protection system plays a critical role in detecting and isolating faults in the HVDC network. These systems employ protective relays, circuit breakers (CBs) and fast-acting devices to swiftly isolate faulty sections, preventing cascading failures. Fault handling mechanisms are essential for ensuring the reliability and resilience of HVDC systems.
- **Communication and monitoring:** Control and protection systems rely on robust communication networks to exchange real-time information between converter stations. Sensors and monitoring devices continuously collect performance data of the HVDC system. Communication protocols and technologies, such as fibre optics and digital communication, facilitate accurate and low latency data exchange.
- **Cybersecurity measures:** As HVDC systems become more interconnected and digitized, cybersecurity measures are crucial to protect

against potential threats. Control and protection systems incorporate cybersecurity features, including secure communication protocols, firewalls and intrusion detection systems, to safeguard the integrity and reliability of the HVDC network.

- **Adaptive control strategies:** Control systems employ adaptive strategies to optimize the performance of HVDC systems under varying conditions. These strategies include predictive control, model-based control and machine learning techniques that enhance the adaptability and efficiency of the control systems in response to changes in load, renewable energy input or grid configuration.

## 3. Switchgear in HVDC applications

The implementation of HVDC technology has marked a new era in power transmission, presenting unique challenges and opportunities for switchgear. As HVDC systems become increasingly integral to modern power grids, the role of switchgear becomes paramount in ensuring the control, protection and reliability of electrical circuits. This section explores the specialized considerations and components that define switchgear in the context of HVDC applications, highlighting the distinctive features that set it apart from traditional HVAC switchgear. From CBs designed to handle continuous DC currents to the intricate world of disconnectors, surge arresters, and earthing switches, this exploration delves into the critical elements that enable the seamless operation of HVDC systems. Additionally, the section addresses the reliability, maintenance and redundancy aspects crucial for sustaining the performance of HVDC switchgear in the dynamic landscape of modern power transmission. Join us as we navigate the complexities of switchgear in HVDC applications, unlocking the key insights that underpin the efficiency and resilience of HVDC technology in the evolving energy ecosystem.

### 3.1 Traditional AC circuit breakers vs. HVDC circuit breakers

The shift from traditional HVAC systems to the innovative realm of HVDC technology introduces a paradigm shift in the design and functionality of CBs. Understanding the nuanced differences between traditional AC circuit breakers and their specialized HVDC counterparts is crucial for grasping the intricacies of power interruption and protection in these evolving power transmission systems.

The detailed comparison as shown in Table 11.1 emphasizes the nuanced engineering considerations and challenges that differentiate HVDC circuit breakers from their traditional AC counterparts. HVDC circuit breakers, designed for the specific demands of continuous DC operation, play a pivotal role in ensuring the reliability and resilience of HVDC technology in modern power systems.

**TABLE 11.1 Comparison of traditional AC circuit breakers vs. HVDC circuit breakers.**

| Aspect | Traditional AC circuit breakers | HVDC circuit breakers |
|---|---|---|
| Nature of current interruption | Designed for alternating currents characterized by natural zero crossings, allowing interruption at points of minimal current. | Must interrupt continuous DC flow, devoid of natural zero crossings, demanding innovative approaches for interruption. |
| Arcing conditions and sustained current | Interrupt AC with minimal arcing due to zero crossings, with less emphasis on sustained arcing conditions. | Operate under sustained arcing conditions during interruption due to continuous DC currents, necessitating advanced arc-quenching mechanisms and materials. |
| Switching speed and response time | Suited for standard switching speeds typical in AC systems, with response times adapted to AC fault characteristics. | Demand faster switching speeds and rapid response times to swiftly intervene during fault conditions in dynamic HVDC environments, ensuring grid stability. |
| Converter station integration | Less integrated within converter stations, as AC systems inherently possess zero crossings that aid in interruption. | Integral components within converter stations, facilitating bidirectional power flow between AC and DC systems, necessitating seamless coordination with converters. |
| Materials and design considerations | Materials optimized for AC operation, with emphasis on interrupting at zero crossings and managing AC fault characteristics. | Specialized materials, including advanced insulators and arc-extinguishing systems, tailored for continuous DC operation, focussing on addressing unique challenges of DC systems. |
| Fault detection and protection strategies | Standard protective measures for fault detection in AC systems, with a focus on interrupting AC faults effectively. | Enhanced fault detection strategies, incorporating advanced protective relays, monitoring systems, and algorithms to swiftly identify and isolate faults in HVDC networks. |

*Continued*

**TABLE 11.1 Comparison of traditional AC circuit breakers vs. HVDC circuit breakers.—cont'd**

| Aspect | Traditional AC circuit breakers | HVDC circuit breakers |
|---|---|---|
| Energy dissipation and interrupting DC current | AC circuit breakers can rely on natural zero crossings to aid in current interruption, reducing the need for specialized energy dissipation. | HVDC circuit breakers require efficient energy dissipation mechanisms to interrupt continuous DC currents effectively, as natural zero crossings are absent. This necessitates the development of advanced technologies for arc extinguishing and managing energy during interruption. |
| Cooling systems and thermal considerations | Typically designed with cooling systems optimized for AC fault interruption, considering the intermittent nature of AC faults. | Incorporate advanced cooling systems to manage sustained arcing during DC interruption efficiently, addressing thermal challenges associated with continuous DC currents. |
| Reliability under dynamic operating conditions | Performance optimized for steady-state AC conditions, with less emphasis on rapid adjustments to dynamic changes. | Must exhibit high reliability under dynamic operating conditions, with the ability to quickly adapt to changes in power flow, ensuring grid stability in the dynamic environment of HVDC systems. |
| Maintenance challenges and lifecycle considerations | Standard maintenance practices designed for AC systems, with a focus on periodic inspections and testing. | HVDC circuit breakers may require more specialized maintenance procedures due to the continuous DC operation, emphasizing regular monitoring, diagnostics and preventive measures to ensure prolonged reliability and performance. |
| Grid synchronization challenges | In AC systems, synchronization is inherent due to natural zero crossings. | HVDC circuit breakers play a role in managing grid synchronization challenges, especially in systems connecting asynchronous grids or RESs. |

## 3.2 HVDC circuit breaker topologies

HVDC circuit breakers are instrumental in ensuring the reliability and safety of DC grid during faults. HVDC circuit breakers come in two primary forms:

electromechanical and solid-state, each operating according to distinct principles and serving specific applications. Inverse current injection-type electromechanical HVDC breakers are most suitable for high voltage and current ratings. In this method, a pre-charged capacitor discharges through an inductor, superimposing a high-frequency inverse current on the input current to create a current zero crossing for fault interruption. Electromechanical breakers exhibit cost parity with AC circuit breakers and are available up to 500 kV with a fault-clearing time of approximately 100 ms.

On the other hand, integrated gate-commutated thyristor (IGCT)-based solid-state HVDC breakers offer rapid interruption in the millisecond range. Their operation involves deactivating the IGCT, which results in voltage increase until a varistor connected in parallel to the thyristor conducts, preventing excessive voltages. Nevertheless, solid-state breakers encounter challenges such as high on-state losses and substantial capital costs. Solid-state CBs typically have ratings around 4 kV and 2 kA.

Electromechanical breaker technology is conventionally established and economically competitive, whereas solid-state breaker provides fast interruption times with challenges in losses and costs. The choice between these two types involves considerations of interruption speed, cost-effectiveness and specific application requirements.

### *3.2.1 Electromechanical HVDC breaker*

In the schematic representation presented in Fig. 11.2, the electromechanical CB comprises three integral components, delineating its operational structure. Firstly, the nominal current path serves as the conduit for DC current during standard operation, with the switch closed to facilitate uninterrupted flow. The commutation path is strategically designed with a switch and a resonant circuit comprising an inductor and a capacitor to produce an inverse current. Lastly,

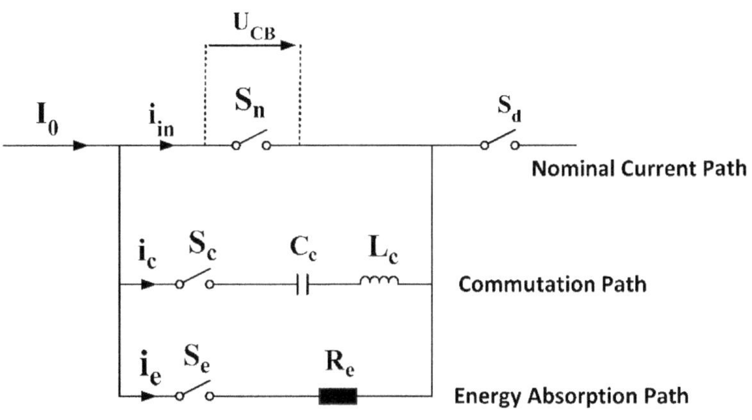


FIGURE 11.2 Electromechanical circuit breaker [2].

the energy absorption path, comprising a switch and a varistor, plays a crucial role in mitigating excess energy during fault conditions.

The commutation path introduces a series resonance, inducing current oscillations between the nominal and commutation paths at the natural frequency $\left({}^{1}/_{LC}\right)$ when interruption is necessary. A zero-crossing event happens when the amplitude of the oscillating current surpasses that of the input current, allowing the switch to disrupt the current flow in the nominal path. Post-interruption, current $(I_o)$ persists, charging the capacitor. When the capacitor voltage surpasses a predetermined value, aligned with the CB's voltage capability, the energy absorption path intervenes, orchestrating a controlled reduction in current.

This fundamental circuit design necessitates further refinements to optimize its efficiency in high-voltage applications. Essential considerations include cost reduction strategies and enhanced utilization of expensive components such as the varistor and capacitor. Additionally, determining the optimum capacitance value holds significance for minimizing interruption time and enhancing overall breaker performance. The interplay of current oscillations and the arc resistance $\left({dU}/_{dt}\right)$ on the switch in the nominal path requires careful management to prevent negative impacts on interruption speed. Moreover, a judiciously chosen C/L (capacitance/inductance) ratio can contribute significantly to maximizing the breaker's interruption performance.

In summary, the electromechanical CB, as illustrated, forms the basis for further advancements aimed at ensuring efficiency and reliability in high-voltage environments. Addressing cost considerations, optimizing component usage and refining key parameters are pivotal aspects that merit continued exploration in the pursuit of enhanced CB technology.

### *3.2.2 Solid-state HVDC breakers*

This research delves into the intricate examination of solid-state CBs, presenting a comprehensive analysis and scrutinizing their economic viability in comparison with other CB topologies. Fig. 11.3 illustrates a distinctive

**FIGURE 11.3** Solid-state circuit breaker [3].

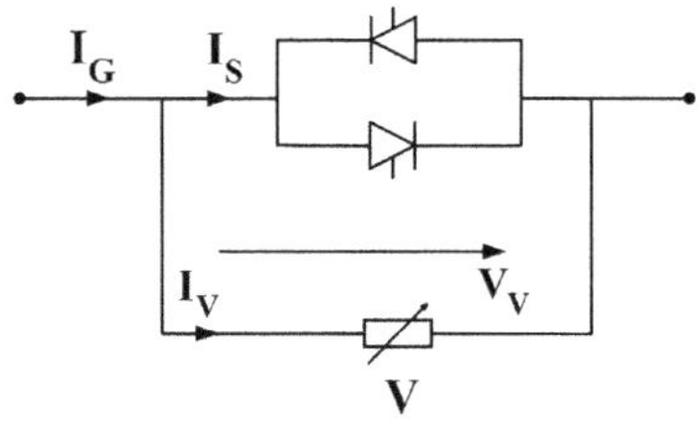

characteristic of solid-state CBs, utilizing gate-commutated thyristors (GCTs) instead of integrated GCTs. This deliberate choice is driven by the paramount concern for minimizing on-state losses within this specific topology.

Under normal operational conditions, where no circuit failure is detected, the current seamlessly traverses through the GCTs. Upon the detection of a fault, a rapid switching-off of semiconductors ensues, inducing a swift voltage surge until the varistor commences conduction. The varistor's design, strategically blocking voltages surpassing the grid voltage, initiates the demagnetization of the line inductance. Notably, this topology, characterized by the absence of parallel connections, employs a total of 14 devices [4].

Solid-state CBs present compelling advantages over alternative topologies. The elimination of mechanical components enhances the responsiveness of these breakers, resulting in markedly reduced turn-off times. Additionally, the relationship between voltage across the inductor and overvoltage allows for a faster demagnetization process, effectively limiting peak currents to low levels. Consequently, for identical grid voltages, power losses per CB are notably diminished in the solid-state configuration.

However, despite these advantageous attributes, the decision to favour solid-state CBs is nuanced due to a significant drawback—high on-state losses. As discerned from the analysis, for low to medium grid voltages, the beneficial characteristics of low turn-off times and limited peak voltage are insufficient to offset the power losses sustained during the on-state. The critical consideration of economic viability unfolds in high-voltage grids, where the substantial power savings per CB outweigh the inherent on-state losses.

In conclusion, the analysis underscores the nuanced trade-offs inherent in selecting the optimal CB topology. While solid-state CBs exhibit compelling advantages in terms of low component cost, swift turn-off times, and limited peak current, their economic viability is contingent upon the specific context of grid voltage. The intricate balance of advantages and disadvantages within the solid-state topology becomes particularly apparent in high-voltage grids, underscoring the need for a judicious assessment of CB options in diverse operational scenarios.

### 3.2.3 Hybrid HVDC circuit breakers

Hybrid circuit breakers (HCBs), amalgamating attributes from miniature circuit breakers (MCBs) and solid-state circuit breakers (SSCBs), aim for a symbiotic balance between minimized operational losses and swift interruption performance. Departing from traditional fault current interruption mechanisms, HCBs improve the performance by executing interruptions within an auxiliary branch running parallel to the main current conduction path. Critical elements of HCBs, which include power electronic devices, are intricately influenced by factors such as voltage and current stresses, on-state voltage drops, failure modes, series voltage equalization effects and the impacts of parallel current sharing.

A meticulous evaluation of the characteristics of insulated gate bipolar transistor (IGBT), integrated gate-commutated thyristor (IGCT) and integrated gate-emitter thyristor (IEGT) reveals the suitability of IGBT and IEGT for high current interruption, while IGCT, with its low on-state voltage and high surge current conduction ability, emerges as the optimal choice for low-voltage, high-current HCBs.

Categorically, HCBs are classified into four main typologies based on commutation principles, responding to diverse operational requirements. These include HCBs based on natural commutation, relying on the arc voltage of a mechanical switch; those based on linear current Source (LCS), incorporating linear current sources; multiport HCBs, integrating multiple ports for adaptability, and HCBs based on capacitor-commutated current disrupters (CCDC), employing auxiliary commutation components.

This refined classification caters to distinct operational demands, with the selection contingent upon factors such as voltage levels, current magnitudes and specific application requirements. The dynamic evolution of HCB technology highlights ongoing research endeavours aimed at achieving heightened performance, reliability and efficiency within electrical power systems.

## 3.3 Other switching devices

In the intricate realm of HVDC applications, a spectrum of additional switching devices augments the functionality and resilience of the overall system, extending beyond the conventional role of CBs. These devices are integral to the control, protection and optimization of HVDC transmission.

### *3.3.1 Disconnectors*

Disconnectors, integral to HVDC applications, stand as pivotal components facilitating the controlled isolation of specific segments within the transmission system. Operating as essential switching devices, disconnectors play a critical role in both routine operational scenarios and maintenance activities, ensuring the safety of personnel and the integrity of the HVDC infrastructure.

The primary function of disconnectors lies in their ability to create a visible gap in the electrical circuit, effectively interrupting the flow of current and isolating designated sections. This capability is particularly crucial during maintenance procedures or in response to faults, providing a means to de-energize specific components for inspection or repair. The design of disconnectors incorporates stringent safety measures, aligning with industry standards to guarantee their effectiveness in creating a reliable break in the electrical connection.

Strategically positioned at key points within the HVDC system, disconnectors offer a controlled and visible means of isolating equipment. Their placement ensures that designated sections can be safely disconnected from the power source, allowing for the execution of maintenance tasks without compromising the overall system integrity. Design considerations for disconnectors encompass factors such as voltage levels, current ratings and the specific operational characteristics of the HVDC network.

The engineering precision behind disconnectors emphasizes responsiveness, reliability and durability. These devices must seamlessly integrate into the broader HVDC infrastructure, operating with precision to meet the demands of the high voltage environment. Ongoing advancements in disconnectors focus on refining their design, incorporating advanced materials and technologies to enhance their performance and align with the evolving requirements of modern HVDC systems. In essence, disconnectors represent a cornerstone in ensuring the operational flexibility, safety and maintenance efficiency of HVDC applications.

### 3.3.2 Surge arrester

Surge arrester, a vital component in HVDC applications, play a critical role in fortifying the system against voltage surges and transient overvoltage. These devices act as protective barriers, diverting excessive voltage away from sensitive components and redirecting it to the ground, preventing potential damage to high-value equipment and ensure the reliability.

The primary function of surge arresters is to swiftly respond to sudden increases in voltage, a common occurrence in power systems due to lightning strikes, switching operations or other unforeseen events of the HVDC system.

Strategically positioned at vulnerable points within the HVDC network, surge arresters provide a robust defence mechanism against voltage fluctuations. Their design considers factors such as the voltage rating of the system, transient energy handling capabilities and response time. Surge arresters consist of advanced materials and technologies, allowing them to respond rapidly to voltage spikes and resume normal operation seamlessly.

The inclusion of surge arresters contributes to the overall resilience and reliability of HVDC applications. As technology advances, research continues to refine the design and performance of surge arresters, ensuring they remain effective guardians against transient disturbances in high-voltage DC systems. In essence, surge arresters exemplify the proactive measures integrated into HVDC networks to safeguard equipment and maintain the integrity of the transmission infrastructure.

### 3.3.3 Earthing switch

Earthing switch constitutes a vital component in HVDC applications, providing essential functions related to grounding and fault management.

These switches play a pivotal role in establishing proper grounding when necessary, contributing to system stability and ensuring the safety of personnel and equipment.

The primary function of earthing switches is to create a controlled grounding path for the HVDC system. During specific operational scenarios or fault conditions, these switches facilitate the safe discharge of fault currents to the ground. By doing so, earthing switches aid in maintaining the overall stability of the HVDC network and prevent potential damage to sensitive components. Strategically positioned at critical points within the HVDC infrastructure, earthing switches must meet stringent performance criteria. Their design considers factors such as the voltage and current ratings of the system, response time and the ability to withstand fault currents. These switches operate in coordination with other protective devices to ensure a swift and effective response to fault conditions.

The precision engineering behind earthing switches emphasizes reliability, responsiveness and the ability to function seamlessly within the high-voltage environment. Ongoing advancements in earthing switch technology focus on refining their design, incorporating advanced materials and enhancing their overall performance to meet the evolving requirements of modern HVDC systems. They operate in coordination with surge arrestors, which protect against over-voltage transients by diverting surge currents. While surge arrestors handle immediate over-voltage events, earthing switches ensure the safe dissipation of residual fault currents. Together, they enhance the reliability and safety of the HVDC network by managing faults and over-voltage events effectively.

## 4. Design and performance considerations

Design and performance are pivotal factors influencing the efficiency, reliability and safety of the entire system. In-depth considerations, particularly in insulation coordination and thermal aspects, are paramount to achieving optimal operation and longevity.

### 4.1 Insulation coordination

Insulation coordination of an HVDC breaker is intricately tied to factors such as grounding methods and the nature of electrical faults. For symmetric monopolar and bipolar HVDC topologies, selecting between low and high impedance grounding systems is essential. In these configurations, ground connections may be made at the secondary side of the transformer or at the midway of one side of a bipolar line. Different fault scenarios, such as line-to-line, line-to-line to ground and line-to-ground faults, are carefully considered during the insulation coordination process. Typically, to prevent ground currents between converters, just one converter is firmly grounded with low impedance. This complex interplay of factors is summarized in Ref. [5].

The converter undergoes a distinct process within this insulation coordination framework. In the pre-fault condition, sub-modules either insert or bypass capacitors, resulting in a combined voltage ($V_{dc}$) across the phase. Upon fault detection, the converter is deactivated, causing current to flow through the sub-module's bottom diode. The AC phase voltage may contribute to the fault, depending on the earthing configuration. To mitigate stress on the anti-parallel diode, protection mechanisms such as activating the bypass thyristor are employed [6].

## 4.2 Thermal considerations

High-voltage circuit breakers (HVCBs) are critical components designed to manage fault current and efficiently carry nominal rated current through closed contacts with minimal losses. In the daily operational routine, these breakers maintain nominal current, resulting in ohmic heating and subsequent temperature rise within these components. Rigorous standards, as outlined by IEC and IEEE, specify maximum temperature rise limit for major components for carrying continuous rated current, preventing damage and ensuring the breaker's reliable performance [7].

A pivotal aspect of ensuring the longevity and safety of HVCBs is the temperature rise test. This test involves passing the rated current through the breaker for an extended period, with temperature sensors strategically placed to monitor the temperature. The significance of the temperature rise test lies in its ability to verify thermal behaviour under various limiting conditions and examine temperature rise under normal and fault scenarios. This test serves as a crucial preventative measure, contributing to damage prevention, reduced downtime, extended operational life and enhanced safety for both the system and personnel working with HVCBs.

### *4.2.1 Heat dissipation in HVDC switchgear*

Efficient heat dissipation in HVDC switchgear plays a critical role in maintaining optimal terminal temperature, particularly for the terminals. Keeping terminals as cool as possible, both during switching processes and when closed and conducting, is essential to extend their lifespan and prevent welding. The IEC (EN) 60947.1 standard allows a maximum terminal temperature rise of 70°C above ambient temperature.

To effectively manage terminal temperature, built-in terminals can be utilized to dissipate heat into surrounding cables and busbars. Maintaining a busbar (or cable) cross-sectional area at least as large as the manufacturer's recommendations, and even larger if feasible, proves beneficial. For instance, the Durakool DEVR60, rated for 600A with 200 $mm^2$ busbars, can handle

1000A for 10 min by increasing to 300 $mm^2$ while remaining well within acceptable terminal temperatures [8].

Table 11.2 in IEC (EN) 60947.1 outlines acceptable temperature rise limits for relay and contactor terminals. These are typically composed of silver-plated or nickel-plated copper or brass and can withstand specified temperature limits. Many a time, the HVDC contactor manufacturers design the terminals a higher factor of safety to withstand temperature more than the $70°C$ limit. Temperature withstanding capacity of various materials used for terminal manufacturing is mentioned in Table 11.2

### 4.2.2 Cooling systems for efficient operation

Ensuring efficient heat dissipation is pivotal for the reliable operation of HVDC switchgear systems. Elevated temperatures, especially in components such as converters and CBs, can significantly impact performance and longevity. To counteract this, robust cooling systems are implemented to maintain optimal operating temperatures and prevent thermal stress on critical components.

Cooling systems are designed to dissipate heat generated during both normal operation and fault conditions. Effective heat dissipation not only prevents overheating but also contributes to the overall efficiency and reliability of the HVDC switchgear. Various cooling techniques are employed, such as forced air cooling, liquid cooling, or a combination of both, depending on the specific requirements and design considerations.

Forced air cooling utilizes fans or blowers to circulate air around heat-generating components, carrying away excess heat and maintaining a consistent temperature. Liquid cooling, on the other hand, involves the circulation of a cooling liquid, such as water or a specialized coolant, to absorb and transfer heat away from critical components. The choice between these methods depends on factors such as the heat load, system design and environmental conditions.

**TABLE 11.2 Temperature rise limits of terminals.**

| Terminal material | Temperature limits (K) |
|---|---|
| Bare copper | 60 |
| Bare brass | 60 |
| Tin-plated copper or brass | 65 |
| Silver-plated or nickel-plated copper or brass | 70 |

## 5. Reliability and maintenance

The sustained and reliable operation of HVDC systems is a critical aspect of modern power transmission, especially considering the unique challenges and complexities inherent in HVDC technology. The significance of reliability in HVDC systems goes beyond conventional concerns and is paramount for ensuring a stable power supply across extensive distances and interconnected grids.

### 5.1 Importance of reliability in HVDC systems

The distinct characteristics of HVDC systems, such as long-distance electricity transmission and asynchronous grid connections, demand a high level of reliability because of involvement of intricate voltage conversion processes and complex control mechanisms. The failure of any component can lead to operational disruptions leading to financial losses, compromise grid stability and may have cascaded effects on the broader energy infrastructure. Therefore, the emphasis on reliability is a strategic imperative, requiring meticulous system design, deployment of high-quality components and the implementation of proactive maintenance strategies.

### 5.2 Strategies for minimizing downtime

Ensuring continuous HVDC operation involves a comprehensive strategy that addresses both the design phase and ongoing maintenance efforts. Here are key strategies to minimize downtime and enhance reliability:

**Regular monitoring and inspection:** Implementing routine monitoring and inspection procedures allows for the early detection of potential issues. Leveraging advanced technologies for monitoring and conducting visual inspections helps identify and rectify problems before they escalate.

**Predictive maintenance:** Employing predictive maintenance techniques, such as condition monitoring and diagnostics, enables the anticipation and proactive addressing of potential failures. This approach significantly reduces unplanned outages and extends the operational life of critical components.

**Redundancy and backup systems:** Incorporating redundancy and backup systems ensures operational continuity in case of component failures. Having backup converters, CBs and control systems minimizes the impact of single-point failures.

**Training and skill development:** Investing in the training and skill development of personnel involved in HVDC system operation and maintenance enhances their ability to respond effectively to issues, troubleshoot problems and execute timely repairs.

**Adaptive design:** Designing HVDC systems with adaptive features allows them to accommodate evolving grid conditions, environmental changes and demand fluctuations. An adaptive design ensures resilience in dynamic operating environments.

**Data-driven decision-making:** Utilizing data analytics and real-time monitoring facilitates informed decision-making regarding maintenance schedules, component replacements and system upgrades. Data-driven insights enable proactive decision-making and contribute to enhanced reliability.

By integrating these strategies into the operational framework, stakeholders in the energy sector can significantly reduce downtime and enhance the overall reliability of HVDC systems.

## 6. Case studies (notable HVDC projects and their switchgear solutions)

### 6.1 800 kV HVDC substation knee type disconnector design by GE in India

GE's design for the 800 kV HVDC substation in India incorporates two types of HVDC switches:

**Bypass switch (BPS):** Installed across each converter valve group, the BPS facilitates rapid bypassing of a converter group being taken out of service, redirecting current into a valve group being restored to service.

**Fast-acting discharge switch (FADS):** Positioned between one pole and a ground-connected resistance, the FADS remains open until an AC network fault occurs. Upon fault detection and clearance by an AC circuit breaker, the FADS swiftly performs a close–open (CO) operation within 300 ms to discharge trapped charge in the HVDC line before re-closing and re-energizing the poles.

This dual-purpose switching device as shown in Fig. 11.4 ensures efficient operation and fault management in the HVDC system, enhancing reliability and safety.

### 6.2 ABB's hybrid HVDC circuit breaker

In ABB's hybrid HVDC circuit breaker as shown in Fig. 11.5, the ultra-fast disconnector (UFD) switch plays a pivotal role in revolutionizing the efficiency and reliability of high-voltage DC applications [10]. This innovative switchgear solution merges the superior switching capabilities of power electronics (IGBT) with the low-loss performance of mechanical switchgear.

The UFD, along with an auxiliary commutation switch, forms a mechanical bypass path, ensuring that current flows through the semiconductors only

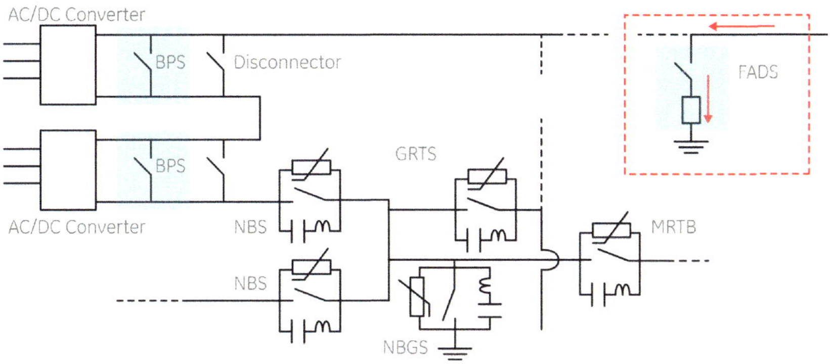


**FIGURE 11.4** Single line diagram of an HVDC transmission scheme using a BPS and FADS [9].

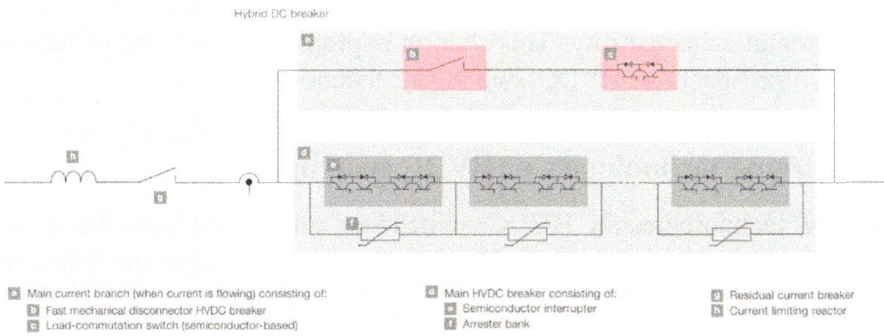


**FIGURE 11.5** Solid-state circuit breaker [10].

when necessary for interruption, thus minimizing losses. Additionally, the load commutation switch enables smooth current transfer from the bypass branch to the main breaker before interruption, allowing the UFD to separate its contacts with virtually no current stress. After the main breaker operates, the UFD is responsible for providing complete dielectric insulation across its contacts, thereby ensuring system integrity [11]. Moreover, designed to endure the full maximum nominal current and equipped for immediate close operation in case of unexpected system subcomponent failures, the UFD stands as a testament to ABB's commitment to pushing the boundaries of HVDC technology.

## 6.3 Siemens's HVDC solutions: Pugalur–Thrissur transmission link

Siemens Limited's groundbreaking project involves the commissioning of India's first HVDC link with voltage-sourced converter (VSC) technology. Connecting Pugalur in Tamil Nadu to Thrissur in Kerala, this significant endeavour comprises two links with a combined capacity of 2000 MW (MW). Implemented in collaboration with Siemens Energy (Germany) and Sumitomo Electric Industries Ltd (Japan), the ±320 kilovolt (kV) HVDC system integrates HVDC Current XLPE Cable with overhead lines, marking a pioneering advancement in India's power transmission infrastructure.

This project, initiated in Feb 2021, plays a pivotal role in addressing the power deficit in the southern region of India while enhancing grid stability. Siemens Limited supplied critical HVDC equipment such as interface transformers and IGBT-based power converters, alongside high- and medium-voltage AC equipment including gas insulated switchgear and air insulated switchgear.

## 7. Future trends and innovations

With the increasing demand for efficient and sustainable energy transmission, the future of HVDC systems is defined by continuous innovation and technological advancement. Emerging trends and innovations are reshaping HVDC switchgear, driving improvements in performance, reliability and environmental sustainability. This section explores key areas of development and future prospects in HVDC switchgear technology.

### 7.1 Emerging technologies in HVDC switchgear

Emerging technologies in HVDC switchgear aim to address the evolving needs of modern power systems. These technologies include the following.

#### *7.1.1 Digital substations*

Utilizing digital substations enhances monitoring, control and automation capabilities, improving the efficiency and reliability of HVDC systems. Digital substations represent a paradigm shift in the management of HVDC systems, offering a host of benefits that elevate the efficiency and reliability of power transmission networks. By leveraging advanced monitoring, control and automation capabilities, digital substations enable real-time data acquisition and analysis, empowering operators to make informed decisions swiftly. This enhanced situational awareness facilitates proactive maintenance and fault detection, minimizing downtime and optimizing system performance. Furthermore, the integration of digital technologies fosters seamless communication and interoperability between various components within the HVDC infrastructure, streamlining operations and enhancing overall grid resilience.

One notable aspect of digital substations is their ability to support advanced cybersecurity measures, safeguarding critical infrastructure against evolving cyberthreats [12]. Through robust encryption protocols, intrusion detection systems and anomaly detection algorithms, digital substations fortify the integrity and confidentiality of data transmitted across the network. Moreover, the adoption of standardized communication protocols ensures compatibility with future technologies and facilitates interoperability with third-party systems, promoting scalability and future-proofing investments in HVDC infrastructure [13]. Overall, the deployment of digital substations marks a significant advancement in the evolution of HVDC systems, paving the way for more efficient, resilient and secure power transmission networks in the digital age.

Digital substations represent a transformative shift in power distribution and transmission infrastructure, leveraging advanced technologies to enhance reliability, efficiency and flexibility:

**Real-time data integration:** Digital substations seamlessly integrate real-time data into the grid, enabling swift diagnosis of issues and reducing downtime. By providing timely insights, they optimize maintenance practices and ensure grid stability, particularly crucial in the context of growing renewable energy integration [14].

**Internet of things (IoT) integration:** Leveraging IoT, digital substations enable continuous monitoring and control, offering comprehensive protection functions and precise measurement of electrical parameters in real time. This connectivity facilitates proactive decision-making, predictive maintenance and improved grid performance, vital for managing the complexities of modern power systems.

**Compactness and reliability:** Digital substations are characterized by their compact design, reliability and safety. By eliminating the need for switchyard panel rooms and enabling remote administration, they reduce project commissioning time and offer greater flexibility in grid management. Moreover, the integration of smart transformers allows for independent voltage regulation, enhancing overall grid efficiency [15].

**Cost efficiency and sustainability:** The deployment of digital substations results in tangible benefits such as lower cost and space requirements, improved productivity and asset reliability. Furthermore, by facilitating remote administration and real-time feedback on power supply parameters, they contribute to environmental sustainability and operational efficiency.

Digital substations represent a significant step towards the modernization of power infrastructure, offering a foundation for smart grid initiatives and enabling utilities to adapt to evolving energy landscapes with agility and resilience.

### *7.1.2 Gas-insulated switchgear*

Gas-insulated switchgear (GIS) substations are gaining traction due to their compact size and high safety index. They offer modularization, lower maintenance and resistance to environmental factors such as vibration and electromagnetic pollution. $SF_6$ gas, traditionally used as an insulating medium, is being replaced with eco-friendly alternatives to mitigate environmental impact.

GIS technology revolutionizes the design and functionality of switchgear, offering several advantages over conventional air-insulated switchgear:

**Compact design:** GIS substations feature a compact, metal-encapsulated design, housing all high-voltage components within gas-filled modules. This compactness reduces the footprint of substations significantly, making GIS ideal for urban areas and indoor installations where space is limited.
**High safety and reliability:** GIS offers high safety indices, lower maintenance requirements and enhanced resistance to environmental factors such as vibration and electromagnetic pollution. The encapsulation of components within gas-filled modules minimizes the risk of electrical faults and ensures reliable operation under diverse conditions.
**Insulation medium:** $SF_6$ gas serves as the insulation medium in GIS, maintaining its molecular and atomic properties even at high voltages. This insulation property allows for closer spacing between active and non-active switchgear parts, reducing equipment size and facilitating installation in confined spaces.
**Environmental considerations:** While $SF_6$ gas offers excellent insulation properties, it is a potent greenhouse gas with a high global warming potential. Efforts are underway to develop eco-friendly alternatives to $SF_6$, addressing environmental concerns associated with its usage. Additionally, GIS technology contributes to environmental sustainability by minimizing land usage and reducing the overall environmental footprint of substations.

GIS technology continues to evolve, with ongoing research focussed on enhancing its performance, reliability and environmental sustainability. As utilities strive to modernize their infrastructure and adapt to changing energy landscapes, GIS remains a cornerstone technology in the transition towards smarter, more resilient power systems.

### *7.1.3 Hybrid switchgear*

Hybrid switchgear integrates the advantages of both air-insulated switchgear (AIS) and gas-insulated switchgear (GIS), offering a balanced solution for various applications:

**Cost efficiency:** Hybrid switchgear strikes a balance between the cost of land and construction expenses by combining AIS and GIS technologies. While initially, the capital cost of hybrid switchgear may be higher than AIS, it offers significant savings in land usage and construction costs, making it a cost-effective solution in the long run.
**Compact design:** By integrating AIS functionality within a gas-insulated enclosure, hybrid switchgear reduces the bay length and overall substation footprint. This compact design optimizes space utilization, making it suitable for applications where space is limited or costly.

**Reliability and maintenance:** Hybrid switchgear inherits the reliability and low maintenance requirements of GIS technology. The encapsulation of components within a gas-filled enclosure enhances resistance to environmental factors and minimizes the risk of electrical faults, ensuring reliable operation under diverse conditions.
**$SF_6$ usage:** Similar to GIS, hybrid switchgear utilizes $SF_6$ gas for insulation, benefiting from its superior dielectric properties. However, efforts are underway to develop eco-friendly alternatives to $SF_6$, addressing environmental concerns associated with its usage. Despite this, $SF_6$-based hybrid switchgear offers operational reliability and performance advantages in demanding environments.

Hybrid switchgear is emerging as a versatile solution, offering a balance between performance, cost and environmental considerations. As utilities seek innovative approaches to modernise their infrastructure and enhance grid reliability, hybrid switchgear presents a compelling option for a wide range of applications in the power sector.

## 7.2 Integration of digital solutions for monitoring and control

The integration of digital solutions is revolutionizing monitoring and control practices in the power sector, particularly in HVDC systems:

**Internet of things (IoT):** IoT technologies enable real-time monitoring and control of HVDC systems by connecting devices and equipment to the internet. This connectivity allows for comprehensive data collection, analysis and remote operation, facilitating proactive maintenance and optimizing grid performance.
**Digital twin technology:** This technology generates virtual replicas of HVDC assets, providing a digital representation of their physical counterparts. These digital twins enable real-time performance monitoring, predictive maintenance and scenario analysis, empowering operators to make informed decisions and optimize asset performance.
**Comprehensive monitoring and protection functions:** Digital solutions offer comprehensive monitoring and protection functions for HVDC systems, including fault detection, isolation and restoration. By leveraging advanced algorithms and analytics, these solutions enhance grid reliability, mitigate risks and ensure seamless operation under varying conditions.
**Data analytics and predictive maintenance:** Advanced data analytics techniques are employed to analyze vast amounts of data generated by HVDC systems. By identifying patterns, trends and anomalies, data analytics enable predictive maintenance strategies, reducing downtime and optimizing asset lifecycle management.

The integration of digital solutions for monitoring and control holds immense potential to transform the power sector, enhancing operational efficiency, reliability and sustainability of HVDC systems. As utilities embrace digitalization and data-driven approaches, they can unlock new opportunities for innovation and optimization in power grid management.

## 7.3 Environmental considerations and sustainability in HVDC systems

Environmental sustainability is a critical aspect of HVDC systems' development and operation, driving innovations and practices aimed at minimizing environmental impact:

**Green converter technologies:** Efforts are underway to develop converter technologies that prioritize environmental sustainability. These green converter technologies aim to reduce energy consumption, emissions and environmental footprint while maximizing efficiency and reliability. By employing eco-friendly materials and improving energy efficiency, green converters contribute to overall sustainability in HVDC systems.

**Alternative insulating mediums:** $SF_6$ gas, traditionally used as an insulating medium in HVDC systems, poses environmental challenges due to its high global warming potential. Research focusses on developing alternative insulating mediums with lower environmental impact, such as dry air or environmentally friendly gases. By replacing $SF_6$ with eco-friendly alternatives, HVDC systems can mitigate greenhouse gas emissions and reduce their environmental footprint.

**Lifecycle assessment (LCA):** LCA methodologies are employed to evaluate the environmental impact of HVDC systems across their entire lifecycle, from manufacturing and installation to operation and decommissioning. By considering factors such as energy consumption, resource utilization and emissions generation, LCAs provide valuable insights for optimizing HVDC system design and operation to minimize environmental footprint and enhance sustainability.

**Renewable energy integration:** HVDC systems play a crucial role in integrating RESs into the grid, facilitating the transition to a cleaner and more sustainable energy mix. By enabling long-distance transmission of renewable energy from remote areas to population centres, HVDC systems contribute to reducing reliance on fossil fuels, lowering greenhouse gas emissions and combating climate change.

Environmental considerations and sustainability are paramount in the design, operation and management of HVDC systems, driving continuous innovation and adoption of eco-friendly practices to minimize environmental impact and ensure a more sustainable energy future.

## 8. Challenges and solutions

The adoption of HVDC transmission systems brings about a myriad of challenges, necessitating innovative solutions for effective implementation and operation. HVDC transmission systems play a pivotal role in modernizing power grids, enabling efficient long-distance power transmission, renewable energy integration and grid stability enhancement. However, the deployment of HVDC technology is not without its challenges. Addressing these challenges requires concerted efforts from industry stakeholders, researchers and policymakers to develop robust solutions that ensure the reliability, efficiency and sustainability of HVDC systems.

### 8.1 Addressing technical challenges

Addressing technical challenges in HVDC transmission systems is essential for ensuring their efficient and reliable operation. From overcoming voltage limitations to enhancing system reliability and efficiency, innovative solutions are being developed to optimize HVDC technology. These solutions aim to push the boundaries of transmission capabilities while maintaining high performance standards, thus facilitating the widespread adoption of HVDC systems in modern power grids.

#### *8.1.1 Overcoming voltage limitations*

HVDC systems often encounter limitations in voltage levels, which can impact their efficiency and long-distance transmission capabilities. To address this challenge, innovative solutions are being developed:

##### 8.1.1.1 Voltage source converters

Voltage source converters (VSCs) revolutionize HVDC transmission systems by addressing voltage limitations and enhancing overall system performance. In traditional HVDC systems, line-commutated converters (LCCs) were primarily used, which had limitations in voltage control and response time. VSC-based HVDC systems offer significant improvements by employing power semiconductor devices such as insulated-gate bipolar transistors (IGBTs) to achieve faster and more precise voltage regulation [16].

These VSCs enable bidirectional power flow, facilitating the integration of RESs and improving grid stability. By efficiently converting AC to DC and vice versa, VSCs ensure smooth power transfer between grids operating at different frequencies or voltages. Their fast response time allows for dynamic adjustment of voltage levels, mitigating voltage fluctuations and enhancing system reliability during transient conditions.

Moreover, VSC-based HVDC systems offer fault ride-through capability, allowing them to continue operating during AC grid faults. This feature ensures uninterrupted power transmission, critical for maintaining grid stability and meeting demand. Additionally, VSCs enable grid operators to optimize power flow and efficiently manage transmission networks, resulting in improved energy efficiency and reduced transmission losses [17].

#### 8.1.1.2 Modular multilevel converters

Modular multilevel converters (MMCs) stand at the forefront of modern HVDC transmission technology, offering a groundbreaking solution to overcome voltage limitations inherent in traditional systems. In HVDC transmission, maintaining voltage stability over long distances poses a significant challenge due to line losses and grid conditions. MMCs revolutionize this landscape by employing a multilevel topology that enables precise voltage regulation and control. By utilizing multiple voltage levels across capacitor banks, MMCs can generate smooth, nearly sinusoidal AC waveforms, thereby minimizing voltage fluctuations and enhancing grid stability, especially in scenarios involving extensive transmission distances.

One of the primary advantages of MMCs lies in their modular architecture, allowing for scalable and fault-tolerant operation. Each submodule within an MMC can be independently controlled, offering flexibility in voltage adjustment and ensuring reliable operation even in the event of submodule failures. This fault-tolerant design enhances system reliability and reduces the risk of voltage instability during transient conditions, addressing key concerns in HVDC transmission networks. Moreover, MMCs facilitate seamless integration into existing AC grids and RESs, supporting bidirectional power flow and enabling efficient utilization of renewable resources. As a result, MMC-based HVDC systems not only overcome voltage limitations but also contribute to the development of sustainable power infrastructure by promoting renewable energy integration and grid resilience.

### *8.1.2 Enhancing reliability and efficiency*

HVDC switchgears are pivotal components in HVDC transmission systems, responsible for the reliable and efficient operation of power networks. Enhancing their reliability and efficiency involves several key strategies:

#### 8.1.2.1 Advanced monitoring and diagnostics

Advanced monitoring and diagnostics systems are indispensable for ensuring the reliability and efficiency of HVDC switchgears. By continuously monitoring critical parameters such as temperature, voltage and insulation condition in real time, these systems enable early detection of abnormalities or potential failures. This proactive approach allows operators to implement timely maintenance actions, preventing costly downtime and ensuring uninterrupted operation of the HVDC transmission system.

Moreover, advanced diagnostic algorithms utilized in these systems enable accurate fault diagnosis and localization within the switchgear components.

This capability facilitates swift remedial actions in the event of a fault, minimizing the impact on system reliability and operational continuity. By swiftly identifying and addressing faults, advanced monitoring and diagnostics systems contribute significantly to maintaining the reliability and availability of HVDC transmission systems.

Furthermore, these systems play a vital role in optimizing the performance of HVDC switchgears. By analyzing operational data and performance metrics, operators can identify areas for improvement and implement corrective measures to enhance system efficiency. Whether through adjusting operating parameters, optimizing cooling systems, or upgrading components, the insights provided by advanced monitoring and diagnostics systems enable operators to maximize the reliability, efficiency and longevity of HVDC switchgears.

#### 8.1.2.2 Advanced monitoring and diagnostics

Advanced monitoring and diagnostics play a crucial role in enhancing the reliability and efficiency of HVDC systems. These systems employ sophisticated sensors and algorithms to continuously monitor various parameters such as voltage levels, current flows and temperature within the HVDC infrastructure. By analyzing this data in real time, potential faults or performance degradation can be detected early, allowing for proactive maintenance interventions. Advanced monitoring and diagnostics enable operators to identify emerging issues before they escalate into critical failures, minimizing downtime and optimizing system performance.

Moreover, these systems provide valuable insights into the health and condition of HVDC components, allowing for predictive maintenance strategies. By detecting subtle changes in operational parameters or trends indicative of impending failures, maintenance activities can be scheduled in advance, reducing the likelihood of unexpected outages and maximizing system availability. Additionally, advanced diagnostic algorithms enable root cause analysis of faults, facilitating efficient troubleshooting and repair processes.

Furthermore, advanced monitoring and diagnostics contribute to the optimization of HVDC system operation. By analyzing operational data and performance trends over time, operators can identify opportunities for efficiency improvements and operational enhancements. For example, by optimizing control algorithms or adjusting operating parameters based on real-time feedback, operators can minimize energy losses, improve system stability and enhance overall efficiency.

## 8.2 Regulatory and standardisation issues

Regulatory and standardisation considerations are pivotal in guaranteeing the safe, reliable and interoperable operation of HVDC switchgear. These issues encompass several key aspects:

### 8.2.1 Safety standards

Regulatory bodies, such as government agencies or industry organizations, establish safety regulations and standards for HVDC switchgear. These regulations cover aspects such as insulation requirements, clearance distances and protection against electrical faults. Manufacturers must adhere to these standards during the design, manufacturing, installation and operation of switchgear to ensure the safety of personnel, equipment and the environment.

When it comes to safety regulations and compliance in practical applications, especially concerning high voltage switchgear, several key aspects are implemented:

**Training and education:** Personnel involved in operating and maintaining switchgear receive comprehensive training on safety protocols and procedures. This training covers aspects such as safe switching operations, isolation techniques, testing procedures and proper grounding practices.
**Adherence to standard operating procedures (SOPs):** Practical implementation involves strictly following established standard operating procedures (SOPs) for all switchgear-related activities. SOPs outline step-by-step instructions for tasks such as switching, isolating, testing and earthing to ensure consistency and safety.
**Use of personal protective equipment (PPE):** Operators and technicians utilize appropriate PPE, including insulated gloves, safety glasses, arc flash suits and face shields, to mitigate the risk of electrical hazards and injuries during switchgear operations.
**Risk assessment and mitigation:** Prior to performing any switchgear-related task, a comprehensive risk assessment is conducted to identify potential hazards and implement necessary controls to mitigate risks. This includes evaluating the voltage levels, presence of hazardous energy and environmental factors.

By implementing these practical measures, organizations ensure compliance with safety regulations and standards while maintaining a safe working environment for personnel involved in HVDC and medium-voltage switchgear operations.

### 8.2.2 Interoperability requirements

Standardization efforts focus on establishing common technical specifications and interfaces to promote interoperability among HVDC switchgear from different manufacturers [18]. This ensures seamless integration and compatibility, facilitating multivendor projects. Here are some practical aspects of interoperability requirements:

**Common communication protocols:** Standardization efforts focus on defining common communication protocols such as IEC 61850 for digital communication in substations. Implementing these protocols ensures that switchgear from different manufacturers can communicate effectively and share vital operational data.

**Unified control systems:** Establishing unified control systems with standardized interfaces enables interoperability among diverse HVDC switchgear. This allows for centralized control and monitoring of equipment regardless of the manufacturer, streamlining operation and maintenance procedures.

**Hardware compatibility:** Standardizing hardware interfaces and specifications ensures compatibility between components and devices from different vendors. This includes connectors, signal levels and voltage ratings, facilitating plug-and-play functionality and reducing integration complexities.

**Testing and certification:** Practical interoperability requirements involve rigorous testing and certification processes to verify compliance with established standards. Testing ensures that switchgear components meet performance criteria and can seamlessly interact with equipment from other manufacturers.

By addressing these practical aspects, standardization efforts promote interoperability among HVDC switchgear, enabling efficient collaboration in multivendor projects and enhancing the reliability and performance of HVDC systems.

### *8.2.3 Grid code compliance*

HVDC switchgear must comply with grid codes, which specify technical requirements and performance criteria to ensure grid stability, reliability and security. Grid code compliance for HVDC switchgear entails adherence to a range of international standards and regulations to ensure safe and reliable operation within power grids worldwide. Adherence to these codes is essential for safe and reliable operation within the power grid. Here are some practical standards commonly used in HVDC switchgear:

**IEC 61850:** This standard specifies communication protocols for substation automation systems. HVDC switchgear often utilizes IEC 61850 for digital communication, enabling efficient monitoring and control of equipment.

**IEC 60694:** HVDC switchgear design and testing often adhere to IEC 60694, which provides specifications for common aspects such as dimensions, performance and testing methods for high-voltage switchgear, and control gear assemblies.

**IEEE C37.123:** This standard outlines requirements for direct current CBs and disconnectors used in HVDC systems. Compliance with IEEE C37.123

ensures the safe interruption of fault currents and isolation of HVDC equipment during maintenance or emergencies.

**CEI/IEC 61869:** HVDC switchgear often incorporates current transformers (CTs) for measuring current levels. CEI/IEC 61869 provides guidelines for the design, performance and testing of CTs, ensuring accurate current measurement in HVDC systems.

**ISO/IEC 27001:** In the context of cybersecurity, ISO/IEC 27001 sets standards for information security management systems (ISMS). HVDC switchgear manufacturers may implement ISO/IEC 27001 to safeguard critical data and protect against cyberthreats, ensuring the integrity and availability of HVDC systems.

**National grid codes:** Additionally, HVDC switchgear must comply with specific national grid codes prescribed by regulatory authorities in different countries. These codes outline technical requirements, performance criteria and operational procedures tailored to the local power grid environment.

By adhering to these standards, HVDC switchgear manufacturers ensure compliance with regulatory requirements, promote interoperability and contribute to the safe and reliable operation of HVDC systems worldwide.

### *8.2.4 Environmental regulations*

Regulatory frameworks address environmental considerations, such as emissions standards and environmental impact assessments, to minimize the environmental footprint of HVDC switchgear installations and ensure sustainability. Practical standards and considerations include the following:

**Emissions standards:** Regulations set limits on emissions from HVDC switchgear operations, particularly sulphur hexafluoride ($SF_6$) gas used in GIS. Standards such as ISO 14001 outline emission reduction strategies, and IEC standards provide guidelines for $SF_6$ handling to minimize greenhouse gas emissions.

**Environmental impact assessments (EIAs):** Before HVDC switchgear installations, EIAs evaluate potential environmental effects. These assessments consider factors such as habitat disruption, noise pollution and visual impacts. Standards such as ISO 14040 and ISO 14044 provide frameworks for conducting comprehensive LCAs to quantify environmental impacts.

**End-of-life management:** Regulations mandate proper disposal or recycling of HVDC switchgear at the end of their operational life to minimize waste and environmental contamination. Standards such as IEC 62271-4 provide guidance on decommissioning, recycling and disposal practices for electrical equipment.

**Energy efficiency:** Efforts to reduce energy consumption in HVDC switchgear contribute to environmental sustainability. Standards such as IEC 61850 facilitate energy-efficient operation through advanced monitoring and control systems, optimizing energy usage while maintaining grid reliability.

By adhering to these standards and regulations, HVDC switchgear manufacturers ensure environmental responsibility and contribute to sustainable energy infrastructure development.

## 9. Conclusion

This chapter has provided an in-depth exploration of various aspects related to HVDC switchgear, focussing on reliability, efficiency, regulatory compliance and environmental sustainability. By examining advanced monitoring and diagnostics, modular multilevel converters (MMC), regulatory and standardization issues, and environmental regulations, this chapter offers valuable insights into the complexities and challenges of HVDC switchgear technology.

From enhancing reliability through advanced monitoring systems to ensuring compliance with grid codes and environmental regulations, HVDC switchgear manufacturers and operators face multifaceted considerations in their quest for safe, efficient and sustainable operation. The discussion on interoperability requirements underscores the importance of standardization efforts to promote seamless integration of multivendor projects, while environmental regulations highlight the imperative of minimizing the environmental footprint of HVDC installations.

Overall, this chapter serves as a comprehensive guide for stakeholders involved in HVDC switchgear design, manufacturing, installation and operation. By embracing technological advancements, adhering to regulatory standards and prioritizing environmental stewardship, the HVDC industry can continue to evolve and thrive in its mission to provide reliable, efficient and environmentally sustainable solutions for future energy infrastructure needs.

## References


[1] A. Kumar, D. Hussain, HVDC (high voltage direct current) transmission system: a review paper, Gyancity J. Eng. Technol. 4 (2) (2018) 1–10.

[2] A. Stan, S. Costinaş, G. Ion, Overview and assessment of HVDC current applications and future trends, Energies 15 (3) (2022) 1193.

[3] C. Meyer, M. Kowal, R.W. De Doncker, Circuit breaker concepts for future high-power DC-applications, in: Fourtieth IAS Annual Meeting. Conference Record of the 2005 Industry Applications Conference, 2005., Hong Kong, China, vol 2, 2005, pp. 860–866.

[4] R. Rodrigues, Y. Du, A. Antoniazzi, P. Cairoli, A review of solid-state circuit breakers, IEEE Trans. Power Electron. 36 (1) (Jan. 2021) 364–377, https://doi.org/10.1109/TPEL.2020.3003358.

[5] C.D. Barker, R.S. Whitehouse, A.G. Adamczyk, M. Boden, Designing Fault Tolerant HVDC Networks with a Limited Need for HVDC Circuit Breaker Operation, CIGRÉ Paris Session, Paris, France, 2014. Tech. Rep. B4-112_2014.

[6] W. Leterme, P. Tielens, S. De Boeck, D. Van Hertem, Overview of grounding and configuration options for meshed HVDC grids, IEEE Trans. Power Deliv. 29 (6) (Dec. 2014) 2467–2475, https://doi.org/10.1109/TPWRD.2014.2331106.

[7] M.T. Dhotre, J. Korbel, X. Ye, J. Ostrowski, S. Kotilainen, M. Kriegel, CFD simulation of temperature rise in high-voltage circuit breakers, IEEE Trans. Power Deliv. 32 (6) (Dec. 2017) 2530–2536, https://doi.org/10.1109/TPWRD.2017.2676841.

[8] https://www.gevernova.com/grid-solutions/products/brochures/disconnector/dc_switch-gear_ gea - 32015-lr. pdf.

[9] https://switchgearcontent.com/2019/04/02/332/ultra-fast-disconnector-switchufd-role-in-abb-hybrid-hvdc-circuit-breaker/.

[10] A. Hassanpoor, J. Häfner, B. Jacobson, Technical assessment of load commutation switch in hybrid hvdc breaker, IEEE Trans. Power Electron. 30 (10) (October 2015) 5393–5400.

[11] https://powerline.net.in/2021/05/25/infrastructure-modernisation.

[12] https://powerline.net.in/2023/02/06/changing-needs-2/.

[13] V.G. Agelidis, G.D. Demetriades, N. Flourentzou, Recent advances in high-voltage direct-current power transmission systems, in: 2006 IEEE International Conference on Industrial Technology, Mumbai, India, 2006, pp. 206–213, https://doi.org/10.1109/ICIT.2006.372391.

[14] https://electricknowhow.com/2023/09/10/what-are-some-of-the-emerging-technologies-in-the-high-voltage-substation-segment/.

[15] J. Hazra, Y. Phulpin, D. Ernst, HVDC control strategies to improve transient stability in interconnected power systems, in: 2009 IEEE Bucharest PowerTech, Bucharest, Romania, 2009, pp. 1–6, https://doi.org/10.1109/PTC.2009.5281816.

[16] F. Schettler, H. Huang, N. Christl, HVDC transmission systems using voltage sourced converters design and applications, in: 2000 Power Engineering Society Summer Meeting (Cat. No.00CH37134), Seattle, WA, USA, vol 2, 2000, pp. 715–720, https://doi.org/10.1109/PESS.2000.867439.

[17] M. Wang, W. Leterme, G. Chaffey, J. Beerten, D. van Hertem, Multivendor interoperability in HVDC grid protection: state-of-the-art and challenges ahead, IET Gener., Transm. Distrib. 15 (15) (2021) 2153–2175.

# Index

'*Note:* Page numbers followed by "f" indicate figures and "t" indicate tables.'

## D

## O



## P



## R



## S

## T



## U



## V



## W



## Z

9780443247187